Dimensions Mat
Workbook PKB

MW01056226

Authors and Reviewers

Pearly Yuen

Tricia Salerno

Jenny Kempe

Allison Coates

Singapore Math Inc.

Published by Singapore Math Inc.

19535 SW 129th Avenue
Tualatin, OR 97062
www.singaporemath.com

Dimensions Math® Workbook Pre-Kindergarten B
ISBN 978-1-947226-15-9

First published 2018
Reprinted 2019, 2020 (twice), 2021, 2022

Printed in China

Acknowledgments

Editing by the Singapore Math Inc. team.
Design and illustration by Cameron Wray with Carli Bartlett.

Contents

Chapter	Exercise	Page

Chapter	Exercise	Page

Blank

Chapter 8 Ordinal Numbers

The children are walking toward the door.
Circle the first child nearest to the door.

Using this page: Have students identify the door and the child nearest to the door as first, then circle the first child.
Concept: Identify the first position.

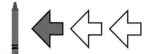

Color the first dog walking down the stairs.

Color the first duck waddling up the hill.

Using this page: Have students identify the front of each line, then count and circle the first dog/duck.
Concept: Identifying the first position.

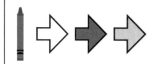

The first animal is yellow.
Color the second one.

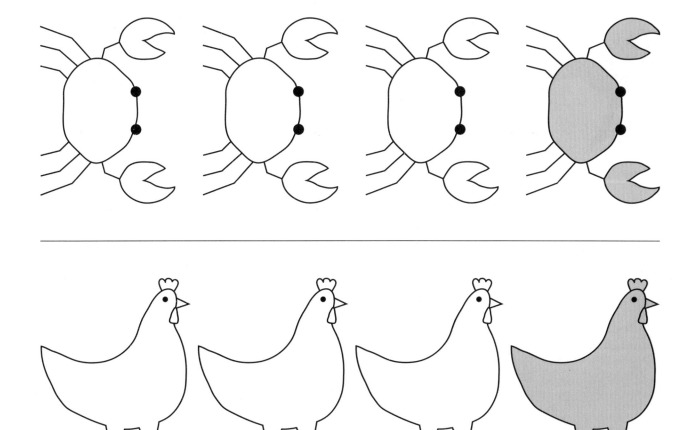

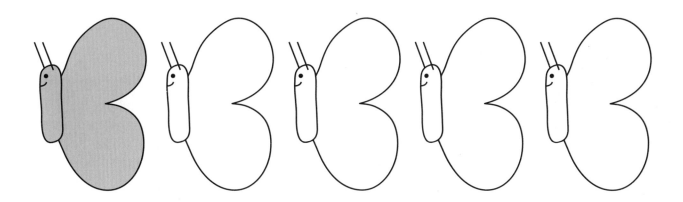

Using this page: Have students identify the yellow animal as the first in line, then count and color the second animal.
Concept: Identifying the second position.

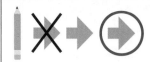

The first animal is circled.
Cross out the third one.

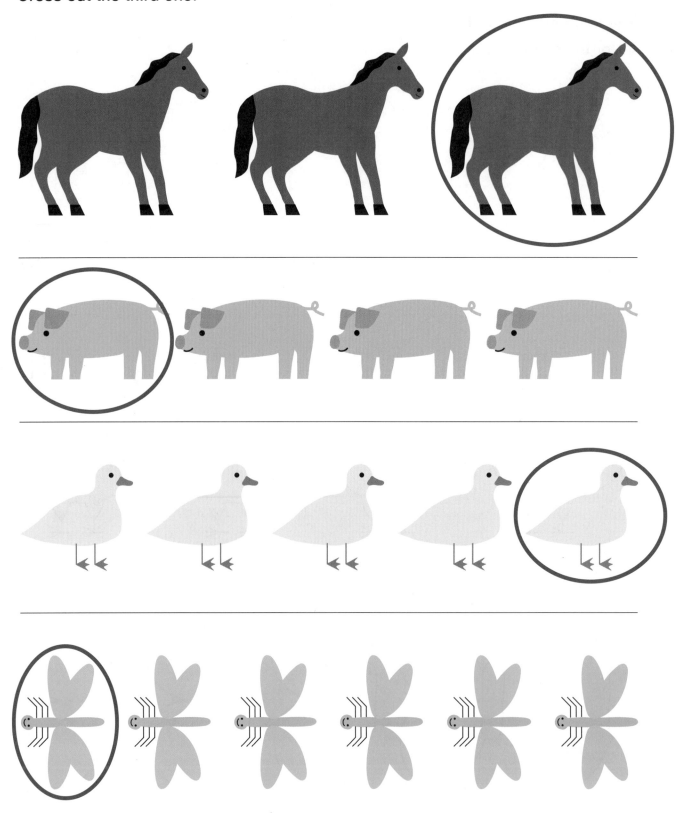

Using this page: Have students identify the circled animal as the first in line, then count and cross out the third one.
Concept: Identifying the third position.

4   8-2 Second and Third

Exercise 3

The first one is small.
Circle the fourth one.
Cross out the fifth one.

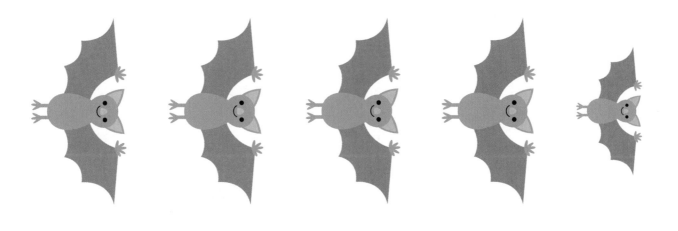

Using this page: Have students identify the smaller animal as the first in line, then circle the fourth animal and cross out the fifth one.
Concept: Identifying the fourth and fifth positions.

8-3 Fourth and Fifth

5

Count from the top and color the fourth one pink.
Count from the top and color the fifth one orange.

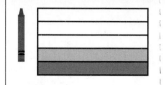

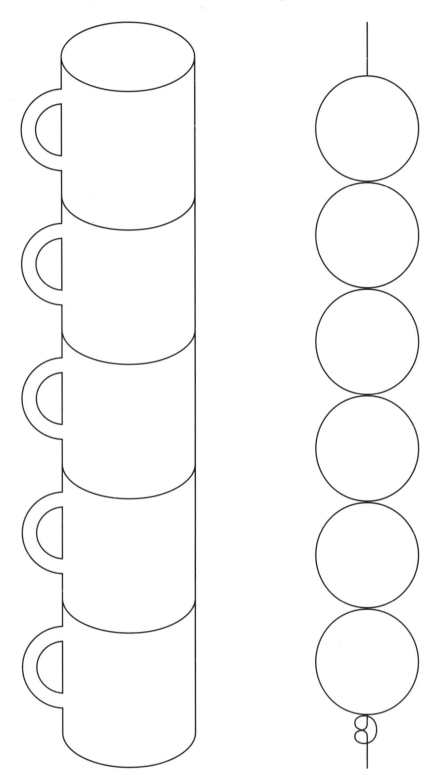

Using this page: Have students identify the top of the stack of mugs, then count from there and color the fourth mug pink and the fifth one orange. Repeat the same process with the string of beads.
Concept: Identifying the fourth and fifth positions.

6   8-3 Fourth and Fifth

Listen and color the ribbons:
- the third ribbon ●
- the fifth ribbon ●
- the second ribbon ●
- the fourth ribbon ●
- the first ribbon ●

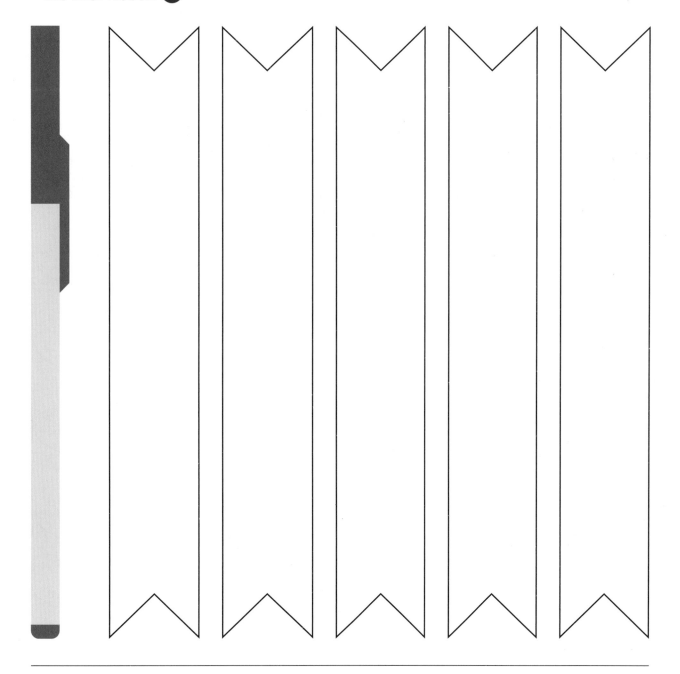

Using this page: Have students identify the pen as the front of the line, then count the ribbons from that direction and color as follows: third ribbon red, fifth ribbon green, second ribbon yellow, fourth ribbon orange, and first ribbon purple.

Listen and color the cake:

The bottom of the cake is the first layer.

- the second layer ●
- the fifth layer ●
- the third layer ●
- the first layer ●
- the fourth layer ●

Using this page: Have students identify the layer that is at the bottom of the cake as first, then counting the layers from there and color as follows: second layer green, fifth layer red, third layer yellow, first layer blue, and fourth layer orange.

8-4 Practice

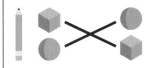

Chapter 9 Shapes and Solids

Match.

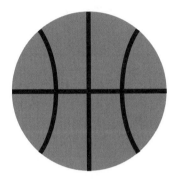

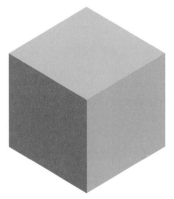

Using this page: Have students match the objects with the solids.
Concept: Identifying same solids.

Cross out the one that does not belong.

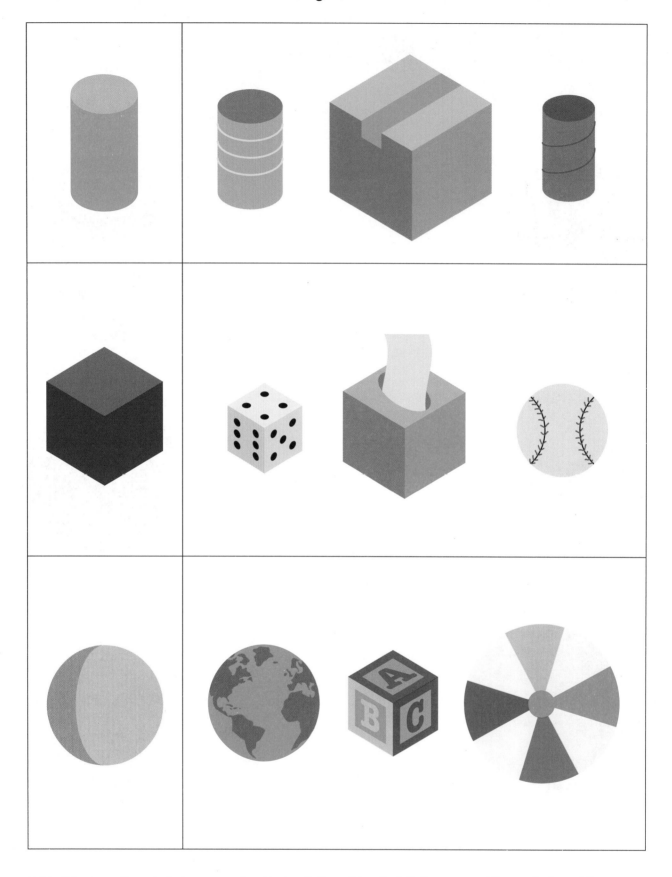

Using this page: Have students compare the objects with the solid on the left, then cross out the one that is not the same.
Concept: Identifying same solids.

10   9-1 Cubes, Cylinders, and Spheres

Color the cubes.

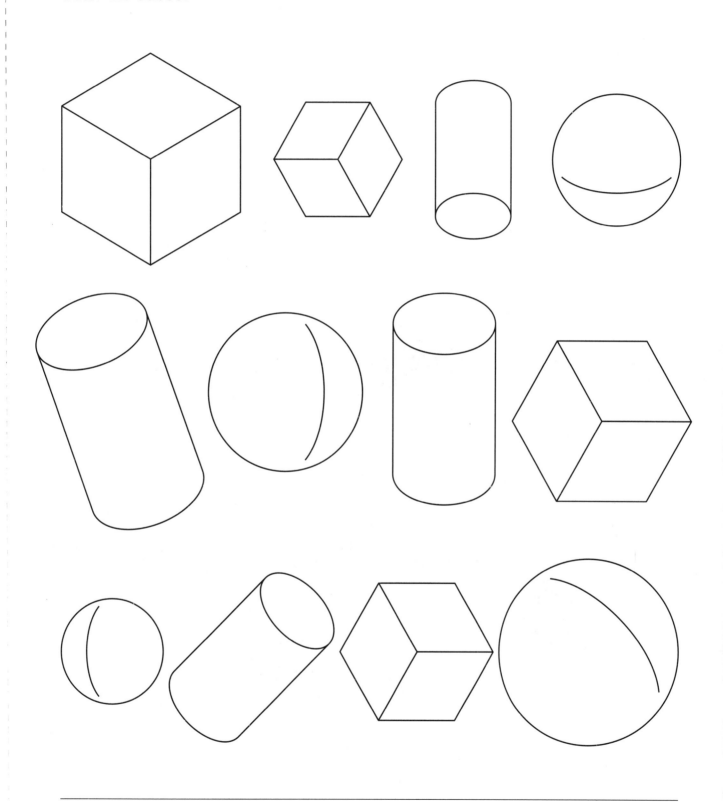

Using this page: Have students identify and color the cubes.
Concept: Identifying cubes.

9-2 Cubes

11

Circle all of the cubes.

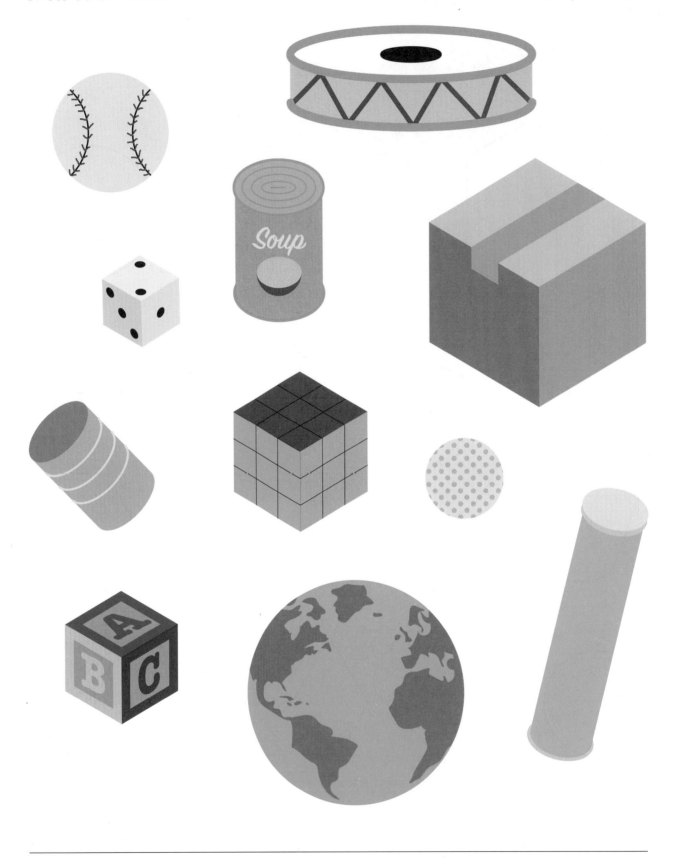

Using this page: Have students identify the cubes and circle them.
Concept: Identifying cubes.

12   9-2 Cubes

Find and color the box with:

- a frog in front of it ⬤
- a frog inside it ⬤
- a frog on it ⬤
- a frog beside it ⬤

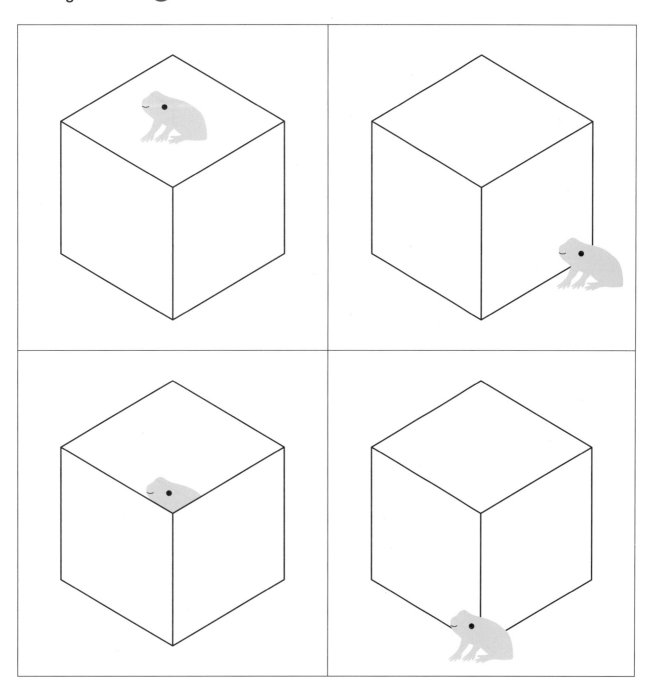

Using this page: As you read each direction to the students, have them identify the correct frog and color each box in the specified color.

Concept: Identifying positions.

Find and color the box or boxes with:

- a frog behind it
- a frog under it ●
- a frog between them ●

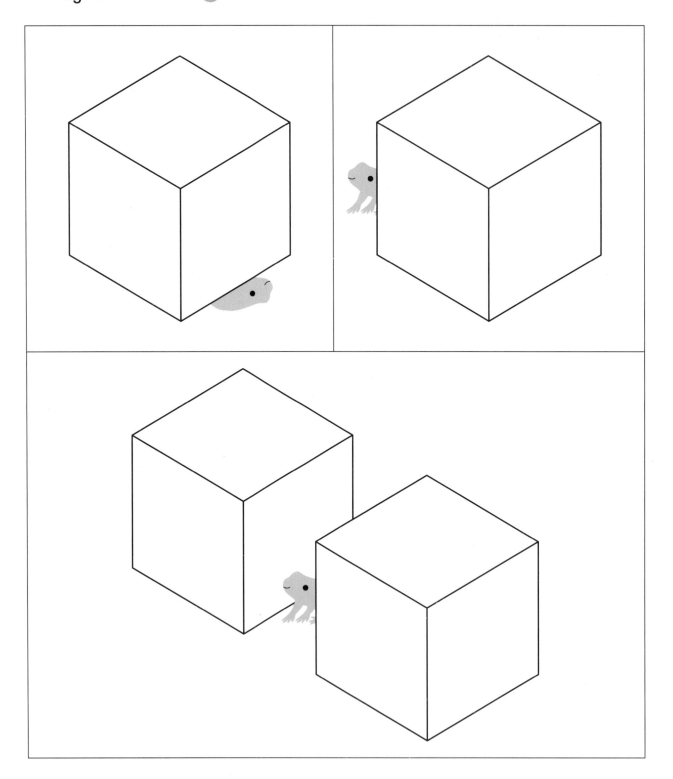

Using this page: As you read each direction to the students, have them identify the correct frog and color the box or boxes in the specified color.
Concept: Identifying positions.

Color the rectangles.

Draw a face on each circle.

Using this page: Have students identify and color the rectangles, and draw a face on each circle.
Concept: Identifying rectangles and circles.

Circle the circles you see.

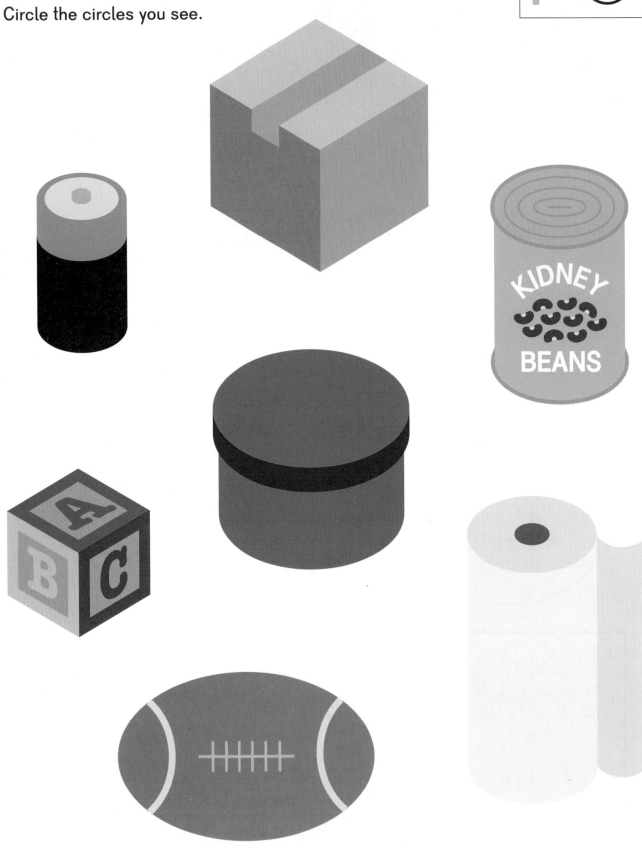

Using this page: Have students identify the solids with a circle for a face/faces and circle.
Concept: Identifying objects with a face in the shape of a circle.

16   9-5 Rectangles and Circles

Color the squares to show how to get the dog to the kennel.

Using this page: Have students identify the squares and color them to show the path from the dog to the kennel.
Concept: Identifying squares.

Match the cubes to the square.

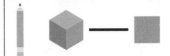

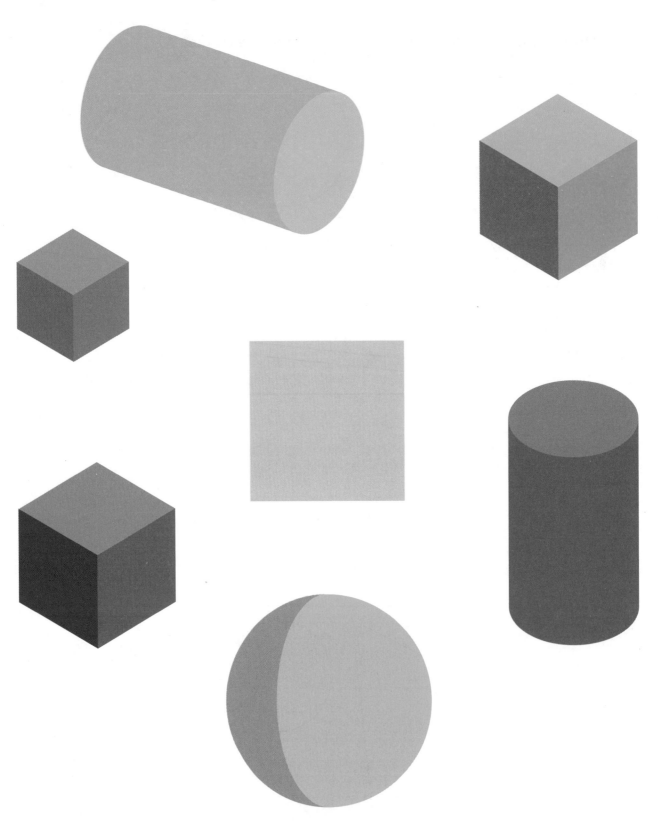

Using this page: Have students identify the solids that have a square as a face and match them to the square.
Concept: Recognize squares as parts of cubes.

18   9-6 Squares

Circle the triangles.

Using this page: Have students identify the triangles and circle them.
Concept: Identifying triangles.

Color the triangles to help the squirrel find the acorn.

Using this page: Have students identify the triangles and color them to show the path from the squirrel to the acorn.
Concept: Identifying triangles.

20   9-7 Triangles

Color the shapes according to the Color Key.

Color Key

Using this page: Have students follow the color key and color the shapes.
Concept: Identifying shapes of different sizes and in different orientations.

Color the shapes according to the Color Key.

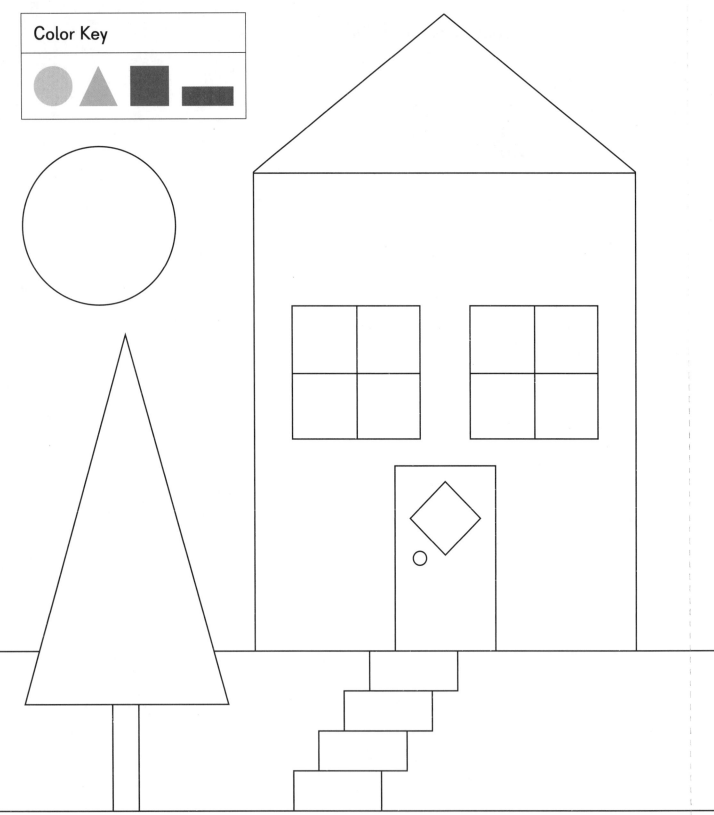

Color Key

Using this page: Have students identify the shapes in the picture and color them according to the color key.
Concept: Identifying shapes of different sizes and in different orientations.

22   9-8 Squares, Circles, Rectangles, and Triangles — Part 1

Exercise 8

Cross out the shapes that are not rectangles.

Cross out the shapes that are not circles.

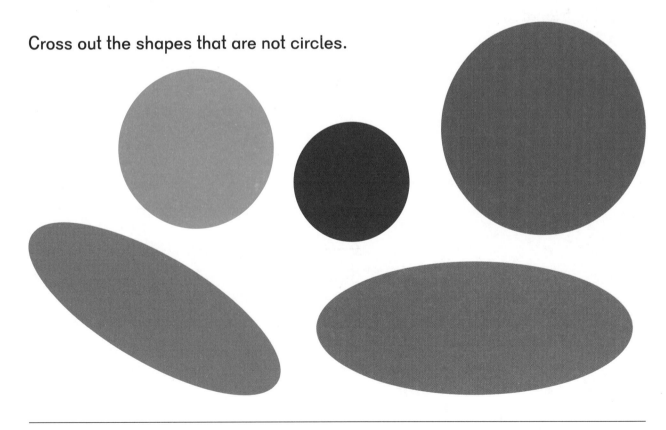

Using this page: Have students identify the shapes that are not rectangles/circles and cross them out.
Concept: Identifying rectangles and circles.

9-9 Squares, Circles, Rectangles, and Triangles — Part 2   23

Trace the squares.

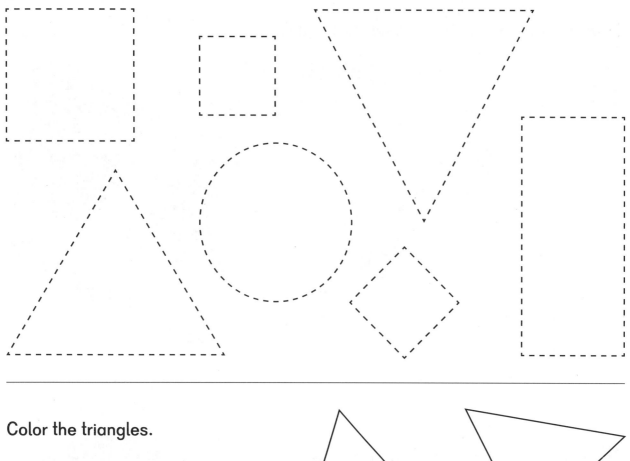

Color the triangles.

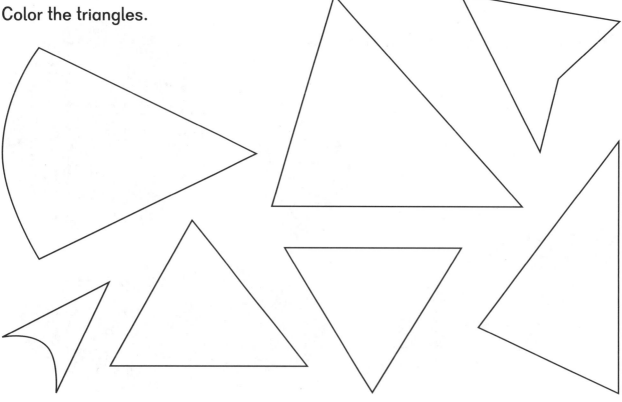

Using this page: Have students identify the squares and triangles and trace/color them according to directions.
Concept: Identifying squares and triangles.

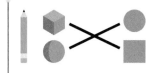

Exercise 9

Match.

Using this page: Have students match the cubes to the square and the cylinders to the circle.

Color the shapes according to the Color Key.

Color Key

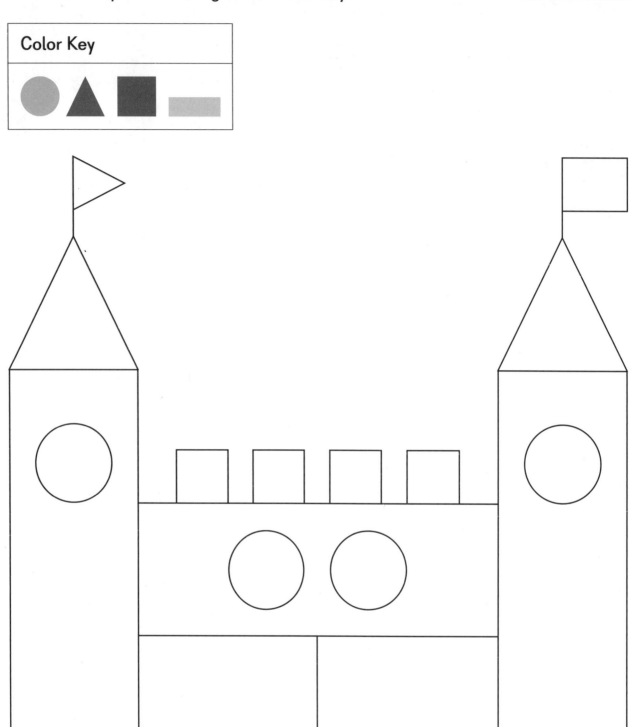

Using this page: Have students identify the shapes in the picture and color them according to the color key.

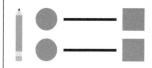

Chapter 10 Compare Sets

Is there a sandwich for each plate?
Match.

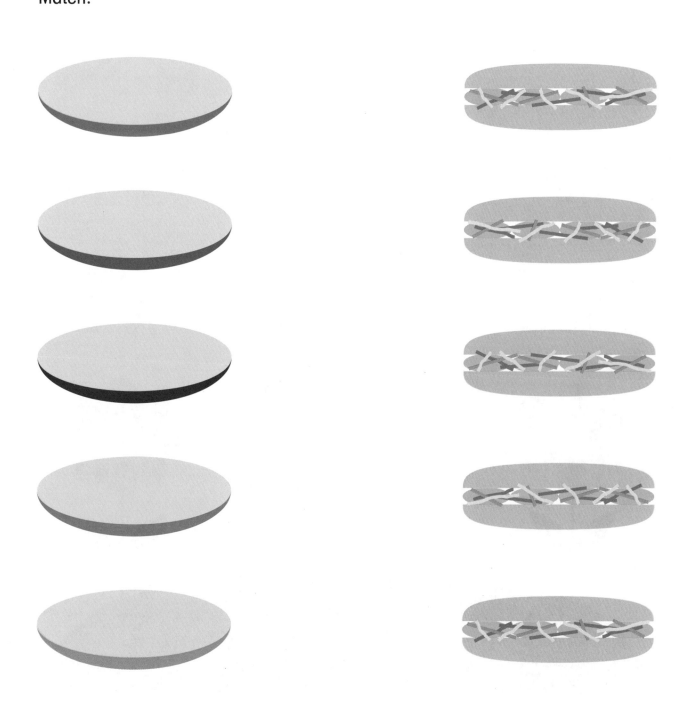

Using this page: Have students match each sandwich to a plate to see if there is one sandwich for each plate.
Concept: Comparing equal sets.

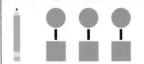

Is there a flower for each butterfly?
Match.

10-1 Match Objects

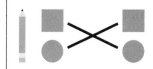

Is there a paintbrush for each paint palette?
Match.

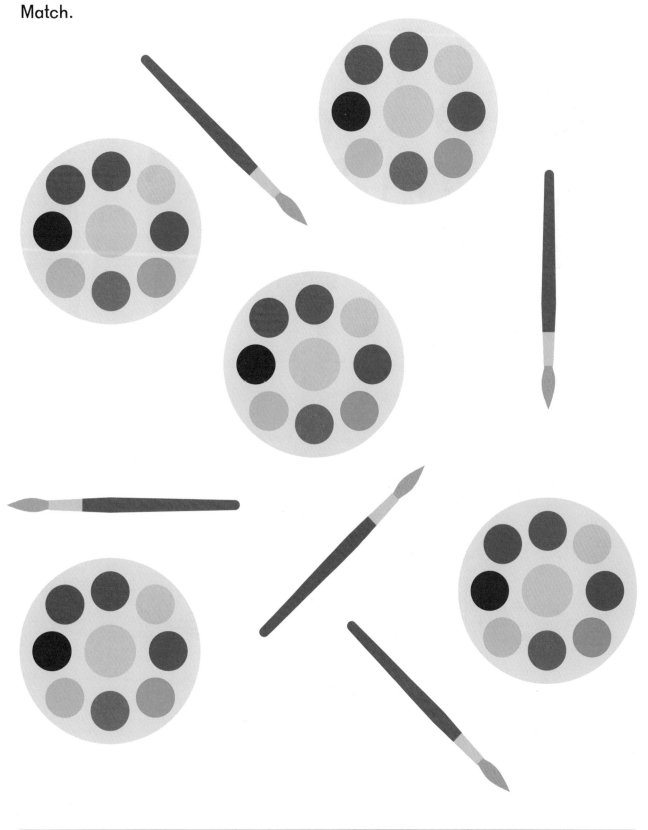

Using this page: Have students match each paintbrush to a paint palette to find out if there is an equal number of paintbrushes and paint palettes.
Concept: Comparing equal sets.

10-1 Match Objects

29

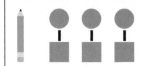

Is there a fish for each child?
Match.

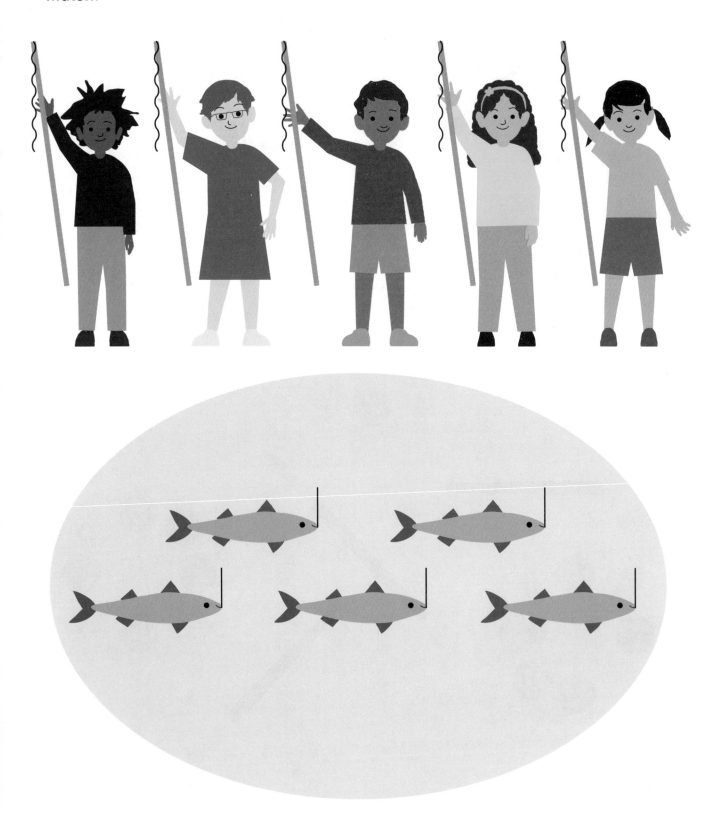

Using this page: Have students extend each fishing line to a hook to see if there is a fish for each child.
Concept: Comparing equal sets.

30   10-1 Match Objects

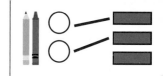

Exercise 2

Match each shirt to a pair of shorts.

Which group has more?

Color that group.

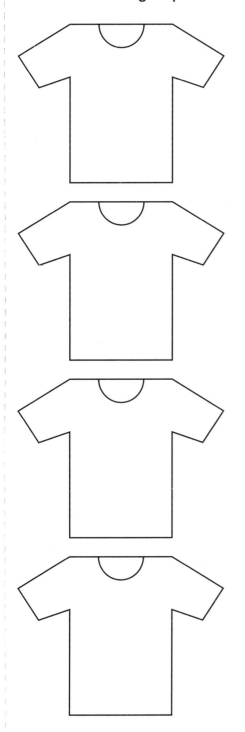

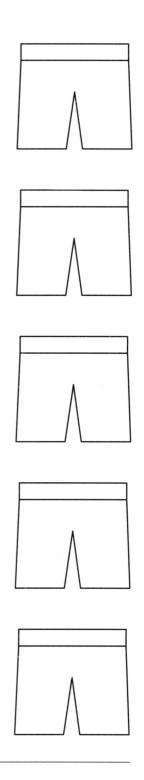

Using this page: Have students match the shirts and shorts to find out which set has more, then color that set.
Concept: Identifying the set that has more.

Which group has more?
Circle it.

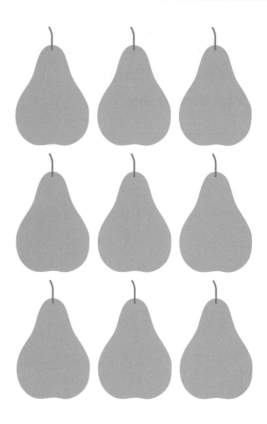

Using this page: Have students compare the number of objects in the two sets, then circle the set that has more.
Concept: Identifying the set that has more.

32   10-2 Which Set Has More?

Exercise 3

Which group has fewer?
Circle it.

Using this page: Have students compare both sets of objects and circle the set that has fewer.
Concept: Identifying the set that has fewer.

Which group has fewer?
Color that group.

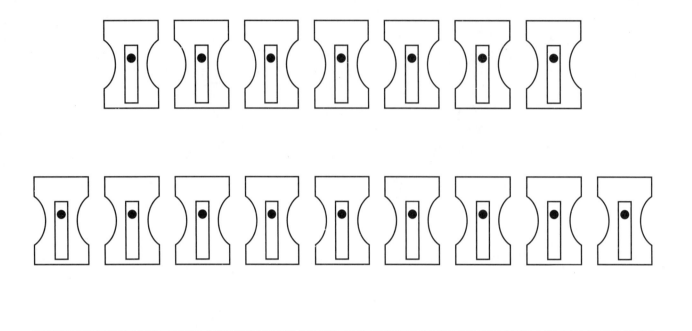

Using this page: Have students compare both sets of objects and color the set that has fewer.
Concept: Identifying the set that has fewer.

Circle the group that has more.

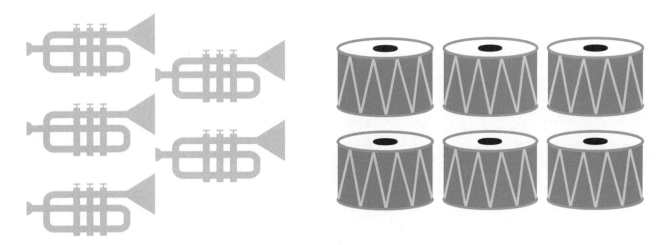

Using this page: Have students compare both sets of objects and circle the set that has more.
Concept: Identifying the set that has more.

10-4 More or Fewer?   35

Are there more 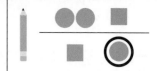 or ?
Circle the answer in the box.

There are more 🌲 .   There are more 🌳 .

Using this page: Have students compare the number of pine trees and oak trees, then circle the answer in the box.
Concept: Identifying the set that has more.

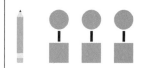

Is there a balloon for each child?
Match.

Using this page: Have students extend the string on each balloon to a child to see if each child has a balloon.

Circle the group that has more.

Circle the group that has fewer.

Using this page: Have students compare both groups of animals and circle the set that has more/fewer.

Chapter 11 Compose and Decompose

Exercise 1

1 horse is eating hay.

3 horses are not eating hay.

How many horses are there altogether?

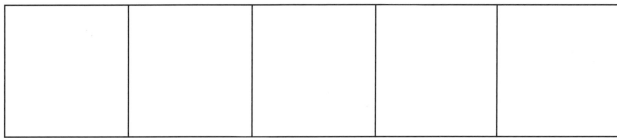

Before using this page: Pre-cut the pictures of horses from cut-outs at back of the book.

Using this page: As you read the number story to students, have them place the same number of horse pictures on the five-frame, then paste the cut-out pictures of horses to show the total.

Concept: Composing four with pictures.

3 polar bears are swimming.

2 polar bears are standing.

How many polar bears are there altogether?

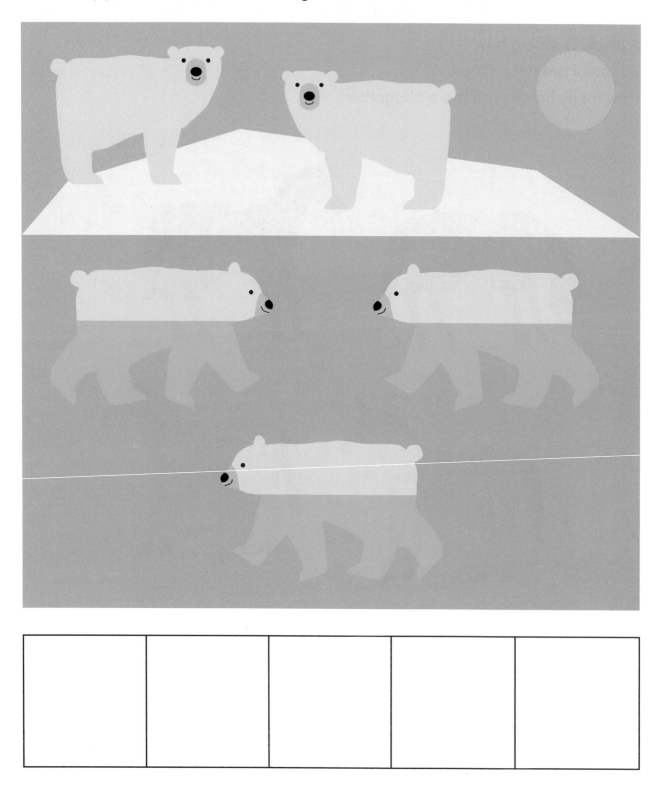

GLUE

GLUE

How many beads are there in all?

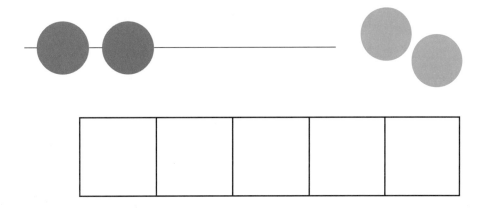

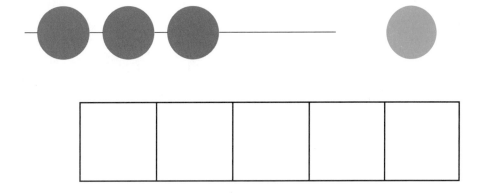

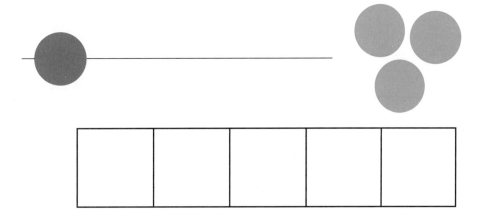

Before using this page: Pre-cut the pictures of beads from cut-outs at back of the book.
Using this page: Have students represent each set of beads by pasting cutouts of the same color and number on the five-frame.
Concept: Composing four with pictures.

How many beads are there in all?

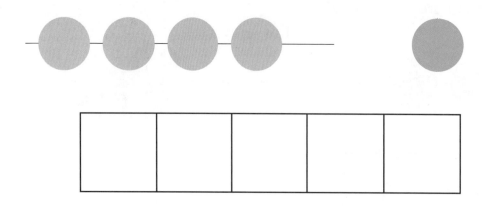

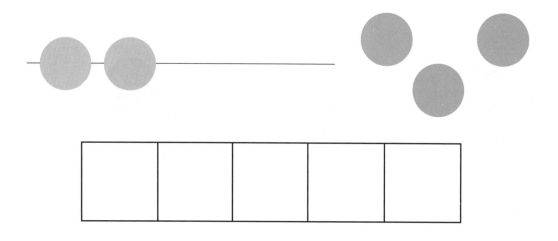

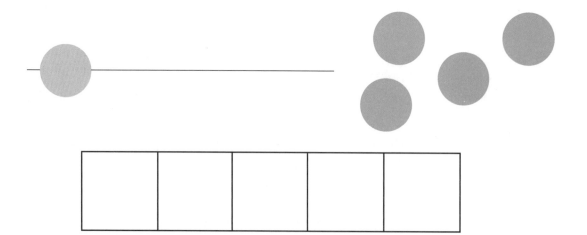

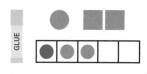

How many are there altogether?

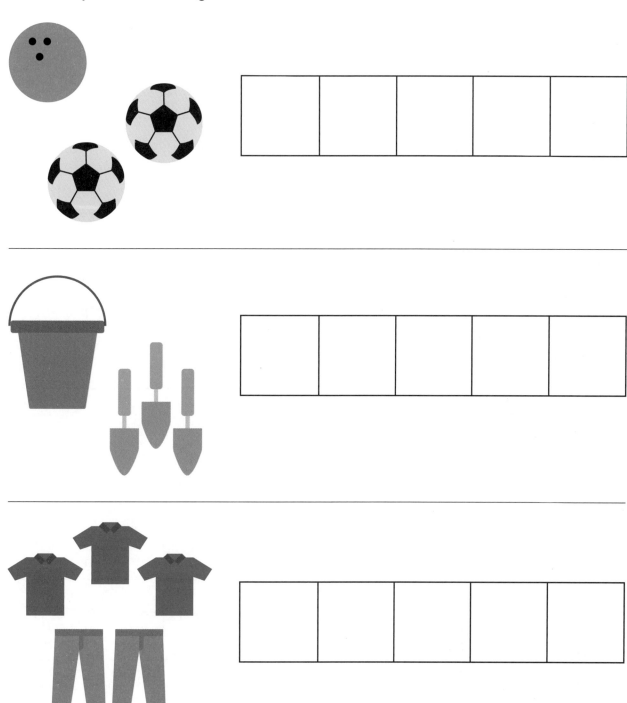

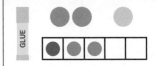

How many are there altogether?

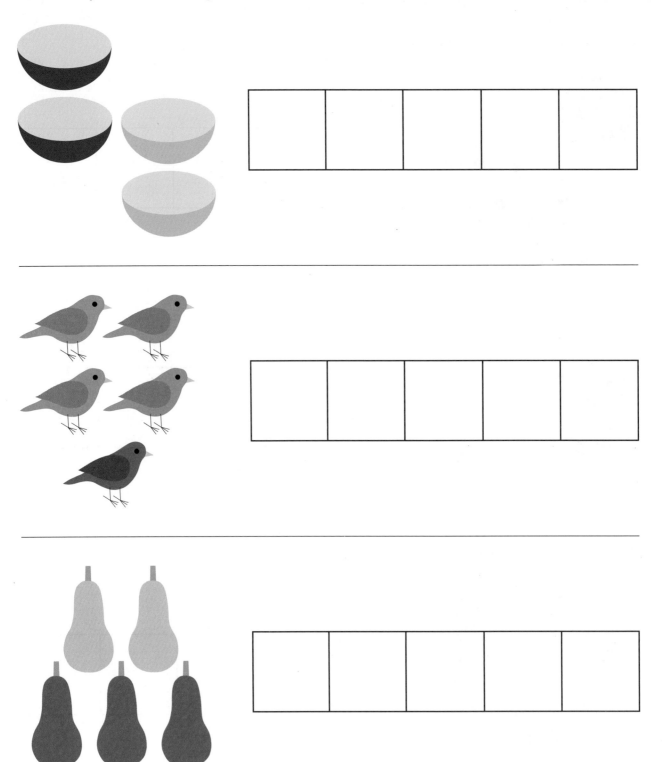

Exercise 4

How many more to match?
Color it.

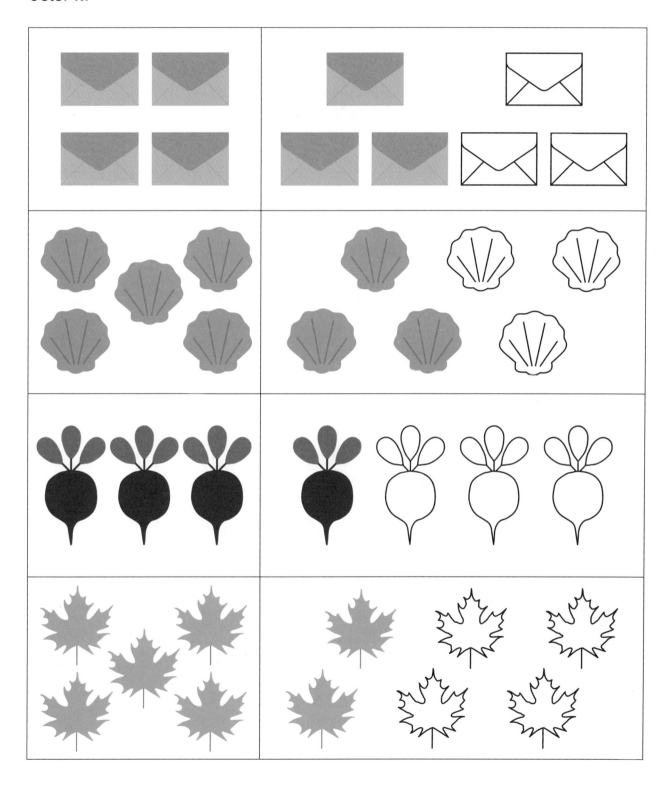

How many more to match what's on the left?
Circle the number.

		3 2 1
		2 5 3
		4 3 1
		4 1 5

Using this page: Have students cross out an object in the whole (shown on the left) for each object in the part (in the center column), to find out how many are in the other part. Then have them circle that numeral in the right column.

Concept: Finding how many in the part not given.

46   11-4 What's the Other Part? — Part 1

Match to make 4.

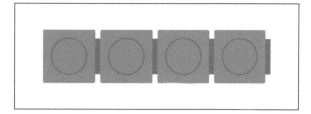

Before using this page: Distribute four counters to each student for the task.
Using this page: Using the top gray cubes as a whole of four, have students put a counter on a gray cube for each purple cube in a part, to find out how many are left. Then match to that number of yellow cubes.
Concept: Finding the other part of four.

Match to make 5.

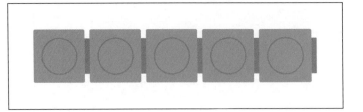

Before using this page: Distribute five counters to each student for the task.
Using this page: Using the top gray cubes as a whole of five, have students put a counter on a gray cube for each purple cube in a part, to find out how many are left. Then match to that number of yellow cubes.
Concept: Finding the other part of five.

48   11-4 What's the Other Part? — Part 1

Each of these ladybugs should have 4 spots altogether on their wings.
Paste the missing spots on the other wing.

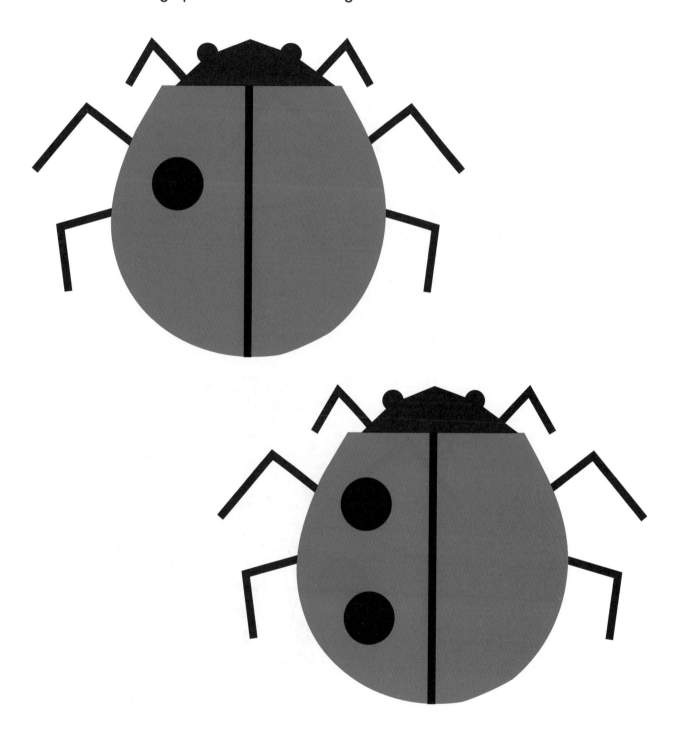

Before using this page: Cut apart eight dot stickers for each student.
Using this page: Have students lay out four dot stickers on the right wing of each ladybug, then take away a sticker for each spot on the left wing. Have them paste the stickers that remain on the right wing.
Concept: Finding the other part of four when one part is given.

Each of these ladybugs should have
5 spots on their wings altogether.
Paste the missing spots on the other wing.

GLUE

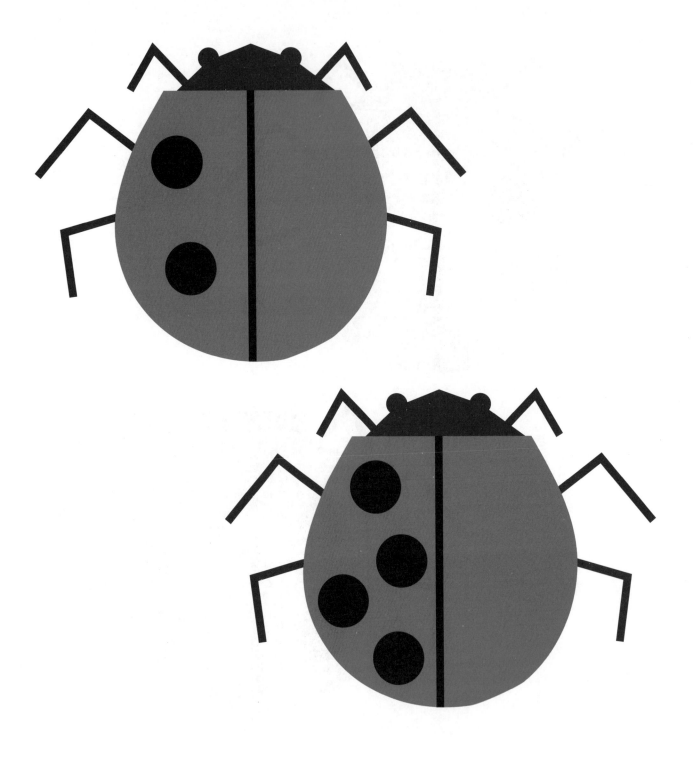

Before using this page: Cut apart 10 dot stickers for each student.
Using this page: Have students lay out five dot stickers on the right wing of each ladybug, then take away a sticker for each
spot on the left wing. Have them paste the stickers that remain on the right wing.
Concept: Finding the other part of five when one part is given.

How many more to make the number on the right?
Color.

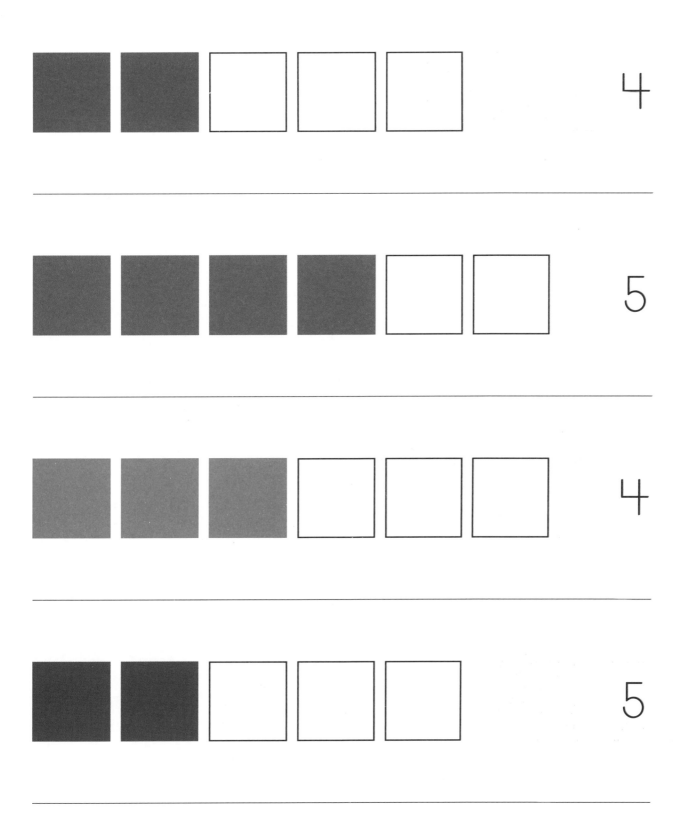

Using this page: Have students lay out the same number of counters as the numeral, then take away a counter for each colored square. Have them color the same number of squares as the numbers of counters left.
Concept: Finding the other part within five when one part is given.

11-5 What's the Other Part? — Part 2   51

How many more to make the number on the right?
Color.

5

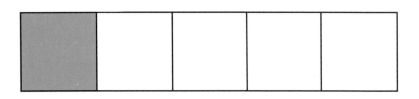

3

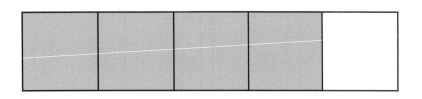

5

4

Using this page: Have students take the same number of counters as the numeral and put them on the five-frame. Then, have them take away the counters on the colored boxes to show the other part, and color in those boxes.
Concept: Finding the other part within five when one part is given.

Exercise 6

How many are there altogether?

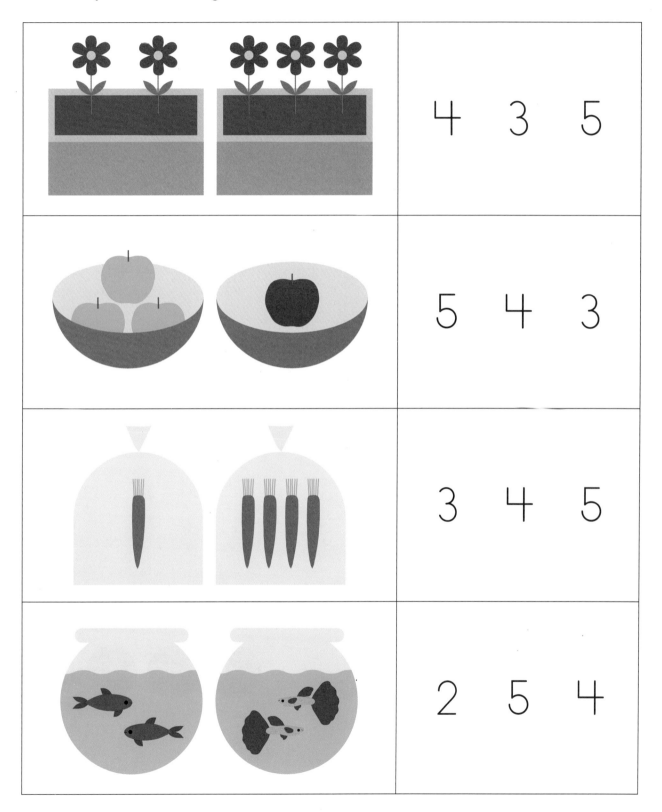

Match to make 5.

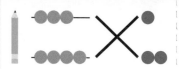

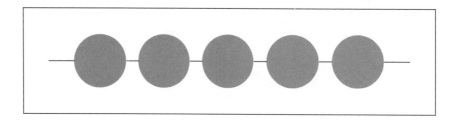

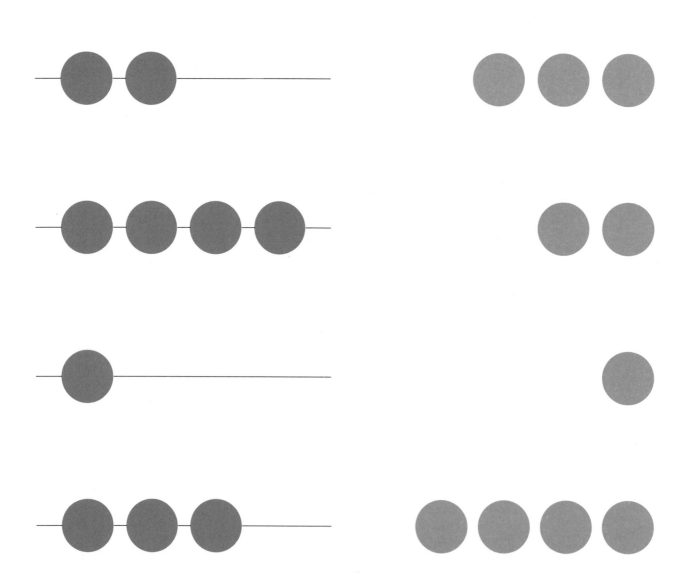

Using this page: Using the string of gray beads as a whole of five, have students put a finger on a bead for each orange bead in a part to find out how many are left. Then have them match that number of green beads.

54   11-6 Practice

Chapter 12 Explore Addition and Subtraction

How many are there altogether?
Color that number.

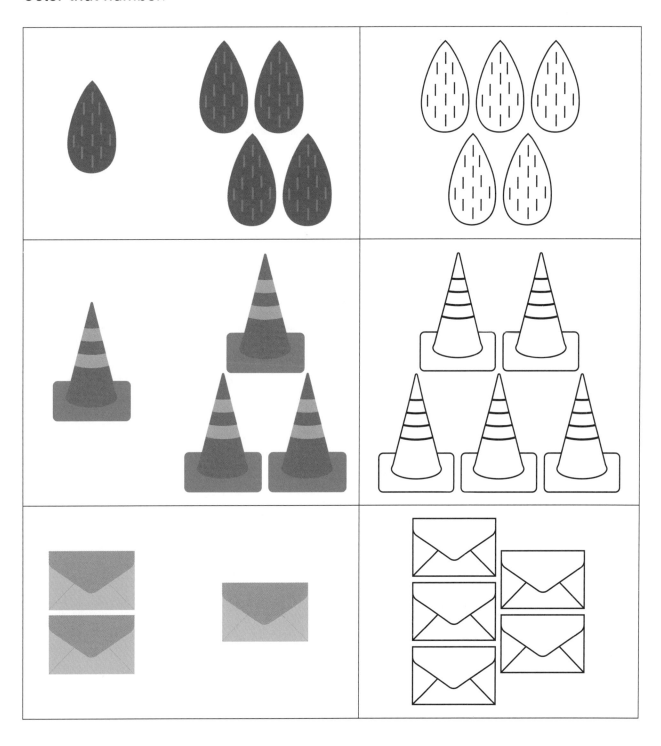

Using this page: Have students add the two parts, then color the objects on the right to show the total.
Concept: Adding to five with pictures.

How many are there altogether?
Match.

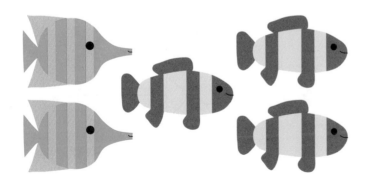

Using this page: Have students add the two parts, then match to that number of counters to show the total.
Concept: Adding to five with counters.

56   12-1 Add to 5 — Part 1

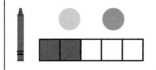

1 duck is in the pond.
3 ducks are getting into the pond.
How many ducks are there in all?

There are 2 pigs on the grass.
There are 3 pigs in the mud.
How many pigs are there in all?

Using this page: As you read the number stories, have students color the same number of boxes for each part on the five-frame to show the total.

Concept: Adding to five with five-frames.

There are 4 oranges in a bowl.

There is 1 pear in a bowl.

How many fruits are there altogether?

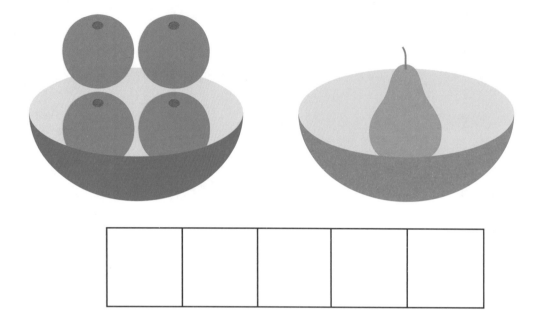

The pet store has 2 brown kennels.

The pet store has 2 green kennels.

How many kennels are there in all?

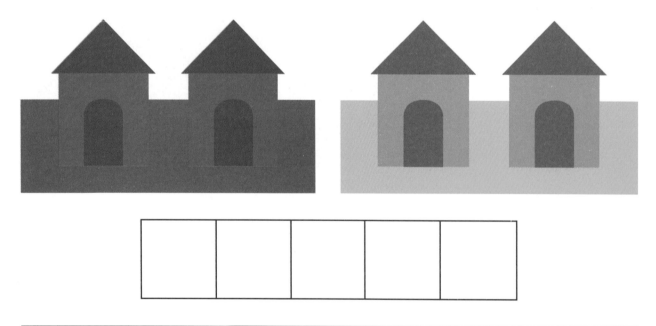

Using this page: As you read the number stories, have students color the same number of boxes for each part on the five-frame to show the total.

Concept: Adding to five with five-frames.

58   12-2 Add to 5 — Part 2

2 dogs are eating.

3 dogs are playing.

How many dogs are there altogether?

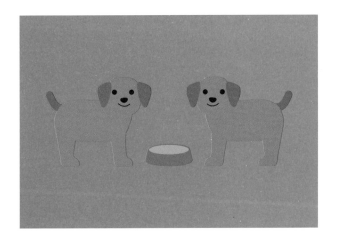

3   2   5

---

1 bird catches a worm.

2 birds are looking for worms.

How many birds are there in all?

2   3   1

Using this page: Read the number stories to students and have them circle the numeral that shows the total.

Concept: Adding to five and number recognition.

The farm has 3 goats.

The farm has 1 cow.

How many animals are there in all?

4   3   1

2 vans are in the parking lot.

2 cars are in the parking lot.

How many vehicles are there in all?

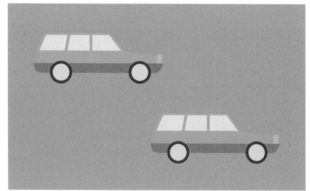

2   4   5

Using this page: Read the number stories to students and have them circle the numeral that shows the total.
Concept: Adding to five and number recognition.

Exercise 3

How many are there altogether?

3   4   2

5   4   1

2   3   5

Using this page: Have students add the two parts, then circle the numeral that shows the whole.
Concept: Adding two parts to make a whole.

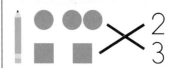

How many are there altogether?
Match.

4

2

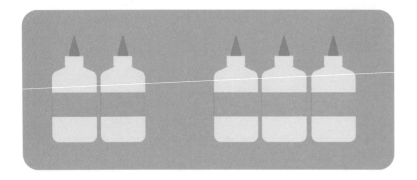

3

5

Using this page: Have students add the two parts, then match that numeral.
Concept: Adding two parts to make a whole.

62   12-3 Two Parts Make a Whole

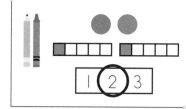

Emma draws 1 red apple.

Then she draws 2 green apples.

How many apples does she draw altogether?

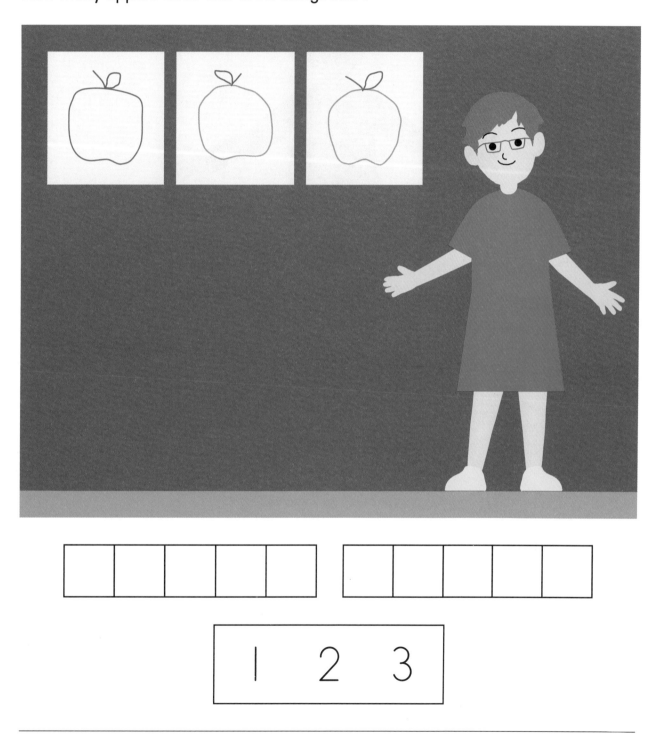

```
1    2    3
```

Using this page: Read the number story to students, then have them color the five-frames for the parts and circle the correct numeral for the whole.

Concept: Adding to five.

Alex makes 3 big paper airplanes.

Then he makes 2 small paper airplanes.

How many paper airplanes does he make in all?

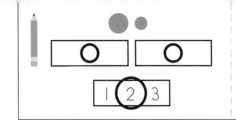

2   3   5

Using this page: Read the number story to students, then have them draw circles for the parts and circle the correct numeral for the whole.
Concept: Adding to five.

There were 3 bats resting in a tree.

1 bat flew away.

How many bats were left in the tree?

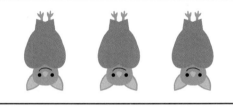 2   1   3

There were 5 bats resting in a tree.

2 bats flew away.

How many bats were left in the tree?

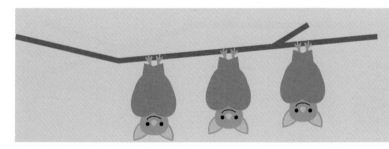

        2   1   3

Using this page: As you read each number story, have students cross out the corresponding number of bats in the box for each bat that flew away to find how many were left, then circle that numeral.

Concept: Subtracting numbers within five with number stories.

There were 4 apples on a tree.

1 apple fell off the tree.

How many apples were left on the tree?

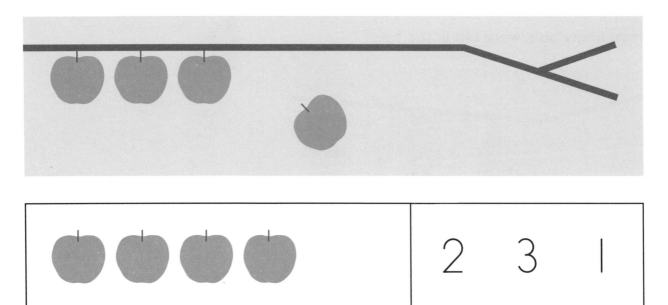

There were 5 apples on a tree.

3 apples fell off.

How many apples were left on the tree?

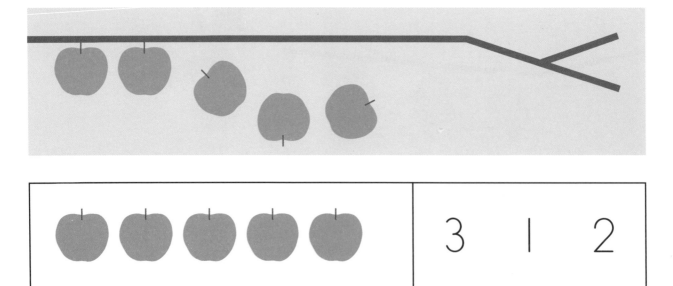

Using this page: As you read each number story, have students cross out the corresponding number of apples in the box for each apple that fell off to find how many were left, then circle that numeral.

Concept: Subtracting numbers within five with number stories.

66   12-5 Subtract Within 5 — Part 1

Exercise 6

4 crabs are on the sand.

2 crabs go to the water.

How many crabs are left on the sand?

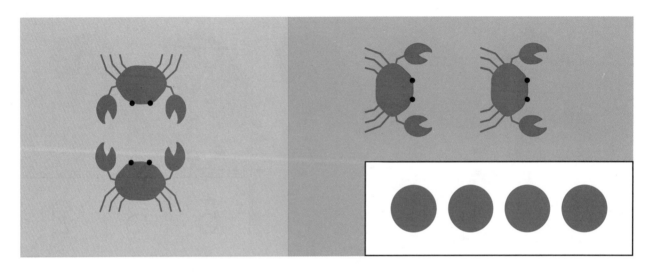

5 snails are on a rock.

1 snail leaves.

How many snails are left on the rock?

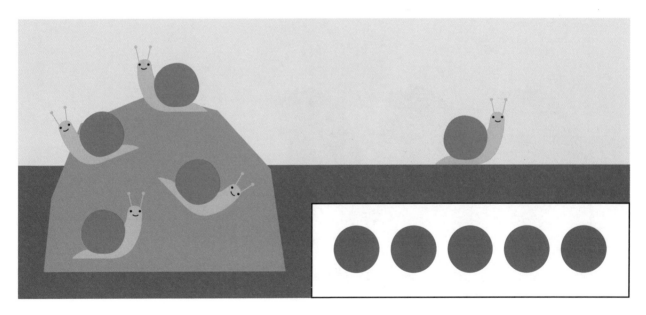

Using this page: As you read each number story, have students cross out the corresponding number of circles in the box for each crab/snail that leaves to find how many are left.

Concept: Subtracting numbers within five with number stories.

5 eagles are on a branch.

2 eagles fly away.

How many eagles are left on the branch?

5   3   2

3 eagles are on a branch.

1 eagle flies away.

How many eagles are left on the branch?

3   1   2

Using this page: As you read each number story, have students cross out the corresponding number of eagles that fly away to find how many are left, then circle that numeral.

Concept: Subtracting numbers within five with number stories.

Circle the number left.

4   3   2

1   3   4

2   4   3

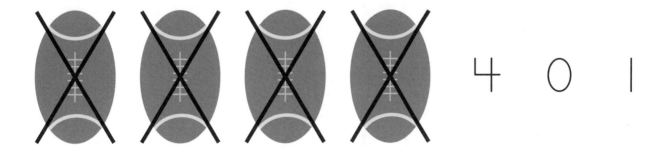

4   0   1

Using this page: Have students find out how many are left, then circle that numeral.
Concept: Subtracting numbers within four.

Circle the number left.

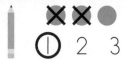

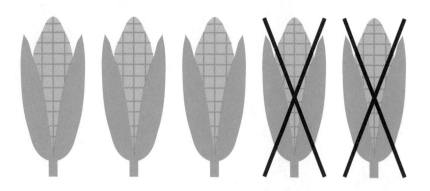

5   3   2

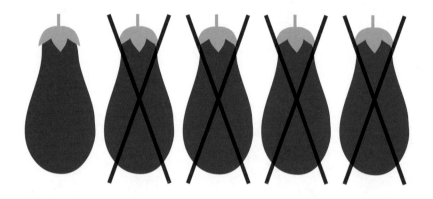

4   5   1

5   1   4

2   5   3

Using this page: Have students find out how many are left, then circle that numeral.
Concept: Subtracting numbers within five.

Exercise 8

How many are there in all?

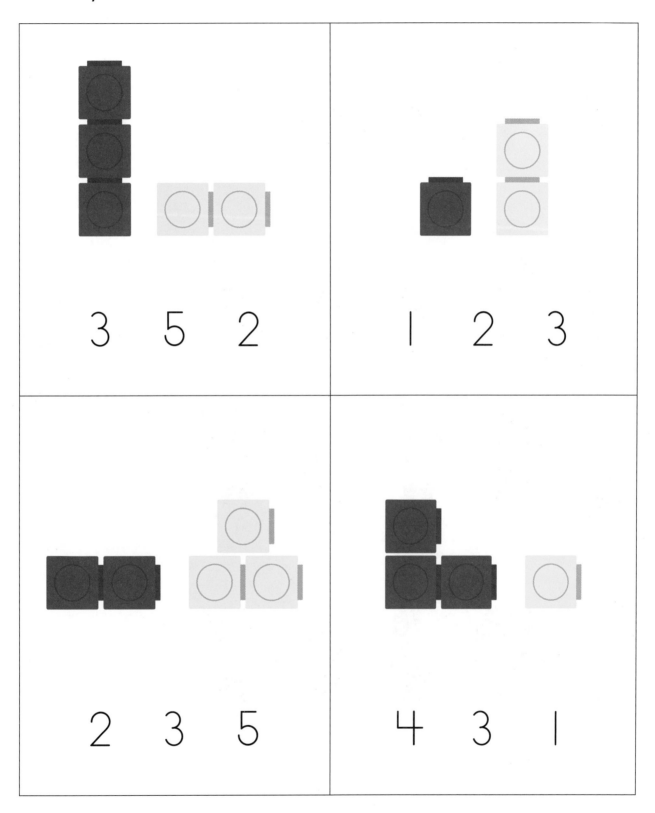

Using this page: Have students add the blue cubes and the yellow cubes, then circle the numeral that shows the total number of cubes.

How many are left?
Match.

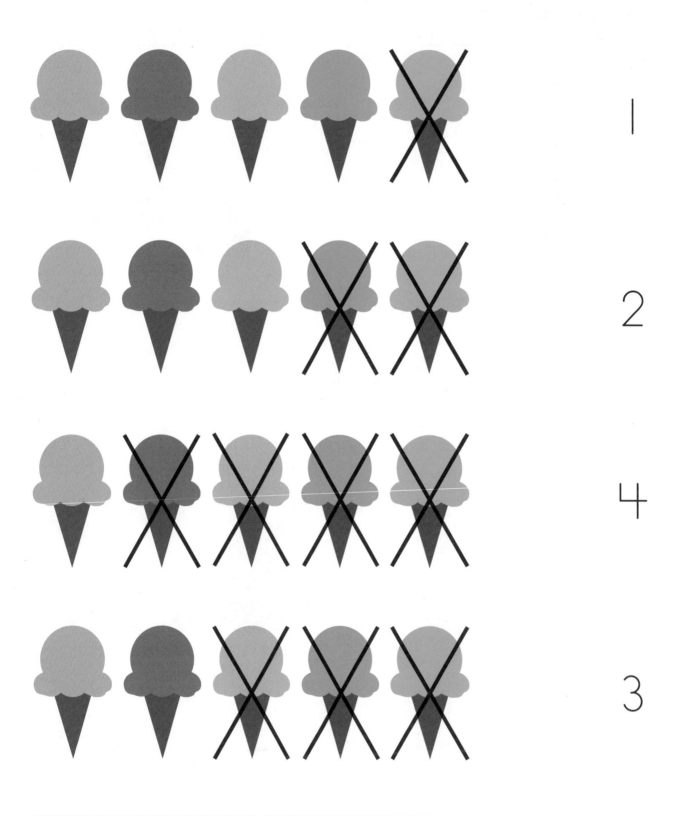

1

2

4

3

12-8 Practice

Using this page: Have students find out how many ice cream cones are left and match to that numeral.

Chapter 13 Cumulative Review

Color the one that is the same.

Using this page: Have students look for the object that matches the colored one and color.

Draw 3 circles for the missing balloons.
Color according to the Color Key.

Color Key

Using this page: Read the directions to students and have them color the parts accordingly.

Circle the big one.

Circle the small one.

Color the small one pink.
Color the big one yellow.

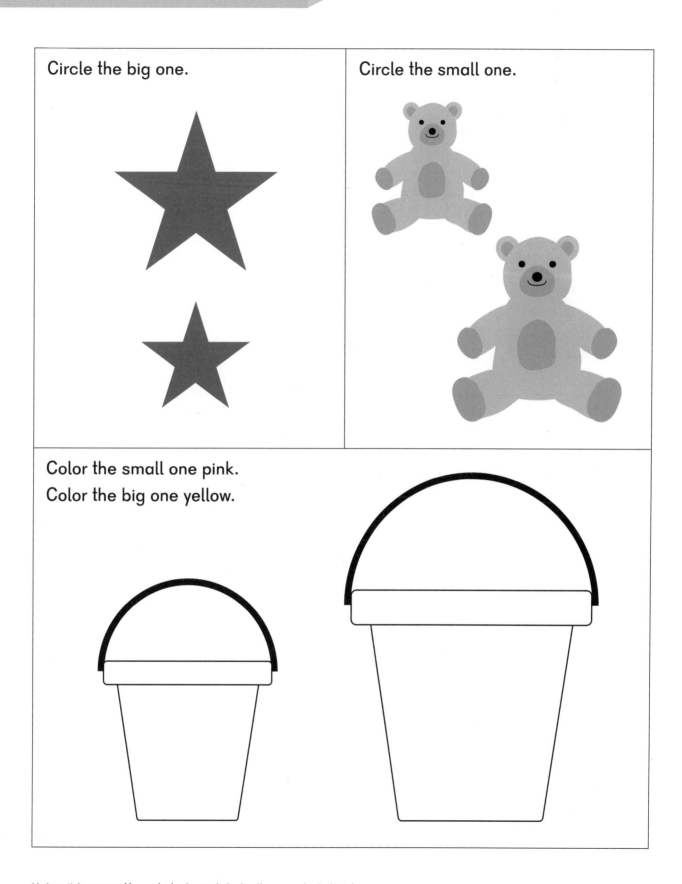

Using this page: Have students circle/color the specified object.

Review 2 Big and Small

75

Color the bigger one.

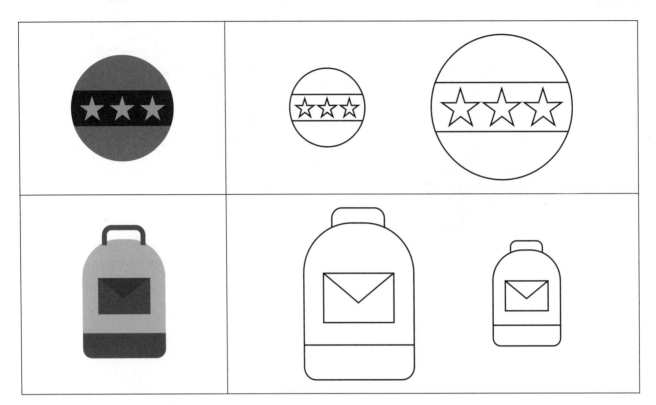

Color the smaller one.

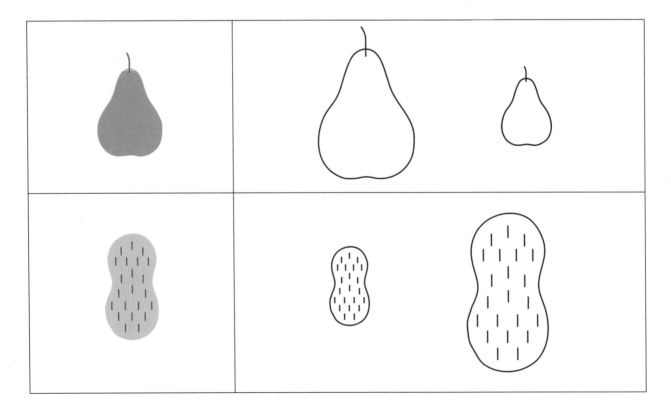

Using this page: Have students color the object that is bigger/smaller than the one on the left.

76   Review 2 Big and Small

Exercise 3

Circle the heaviest object.
Cross out the lightest object.

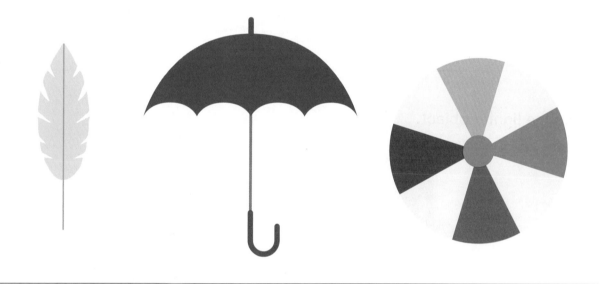

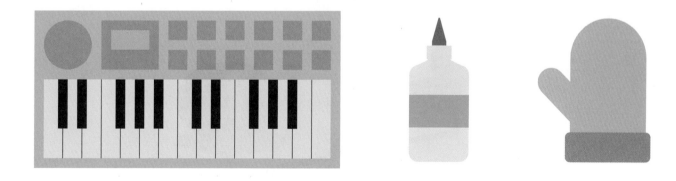

Using this page: Have students circle the heaviest object and cross out the lightest object in each row.

Circle the heavier object.

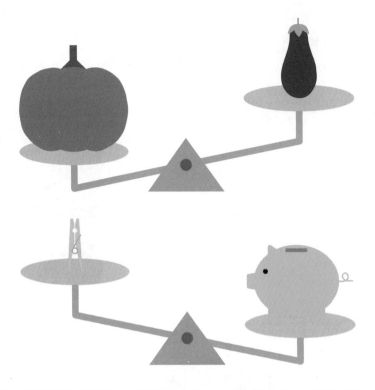

Circle the lighter object.

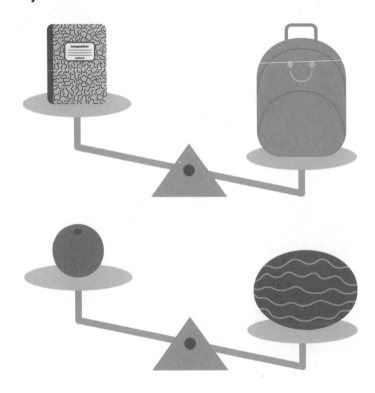

Using this page: Have students circle the heavier/lighter object as specified.

Circle the square that has 5 circles.

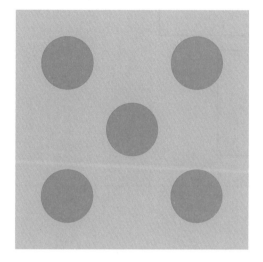

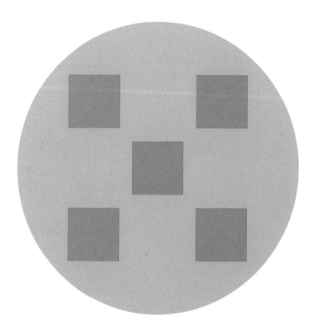

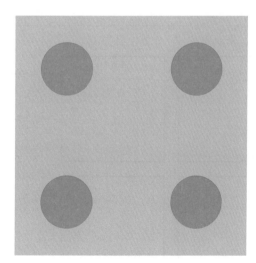

Using this page: Have students identify the correct shapes with the number specified, then circle.

Color the group of 5 squares.

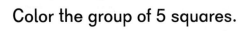

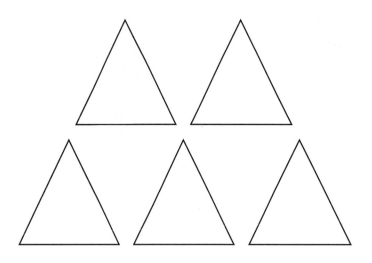

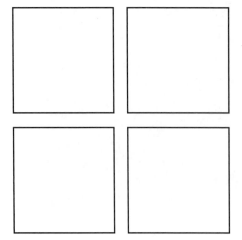

Using this page: Have students count the squares in each group and color as specified.

Color 4 small triangles.

Using this page: Have students identify the triangles that are small and color only four.

Review 5 Count 5 Objects   81

Color 3 small circles.

Cross out 2 big triangles.

Draw a face in the big square.

Using this page: Have students identify the specified shapes and complete the task. For the face, direct students to draw two small circles for the eyes, one small triangle for the nose, and one bigger circle for the mouth.

Review 5 Count 5 Objects

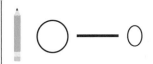

Exercise 6

Match.

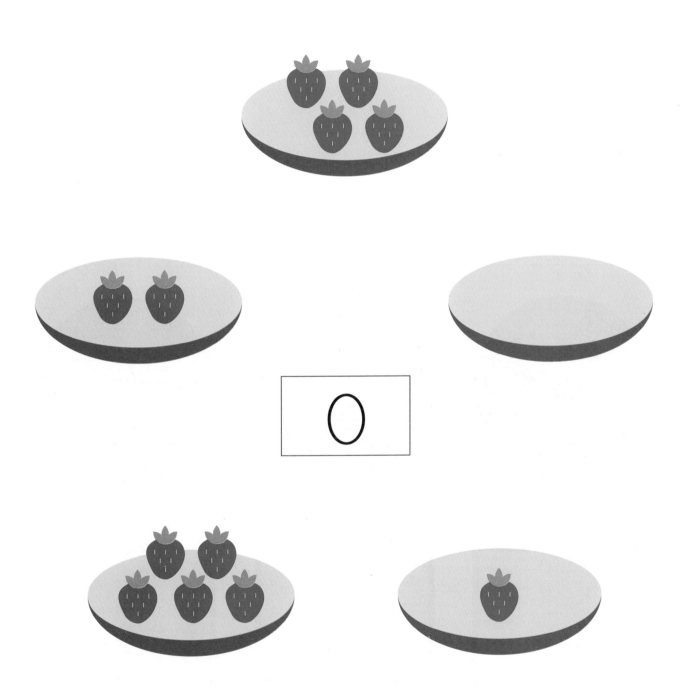

O

Using this page: Have students match the empty plate to the numeral zero.

Circle the one that has zero.

 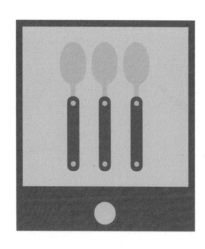 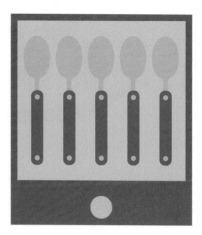

Using this page: Have students circle the set that is empty.

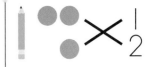

Match.

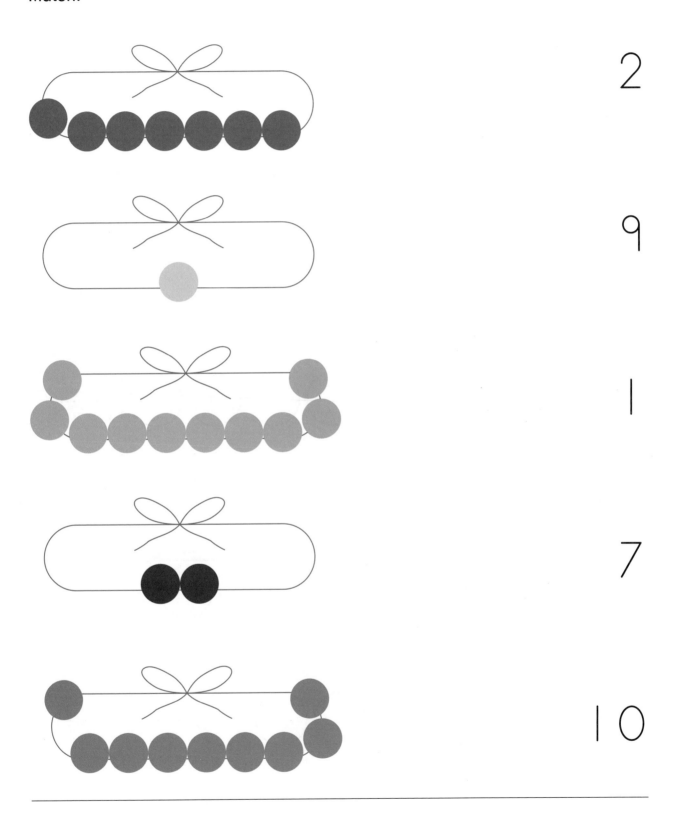

2

9

1

7

10

Using this page: Have students match each set of beads to the correct numeral.

Color the same number of beads.

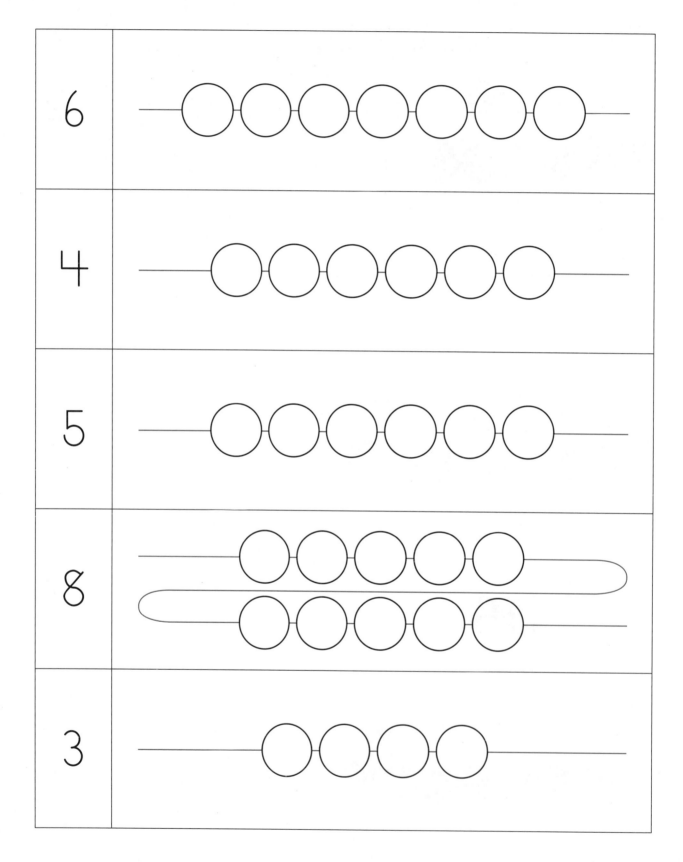

Using this page: Have students color the specified number of beads in each row.

What comes next?

Circle it.

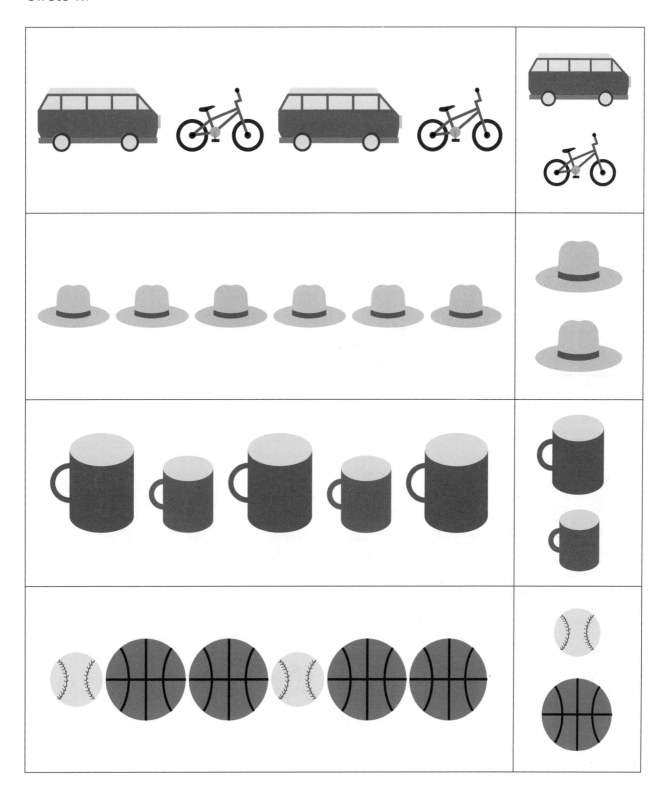

Using this page: Have students look at the pattern in each row and circle the one that comes next.

What comes next?
Color it.

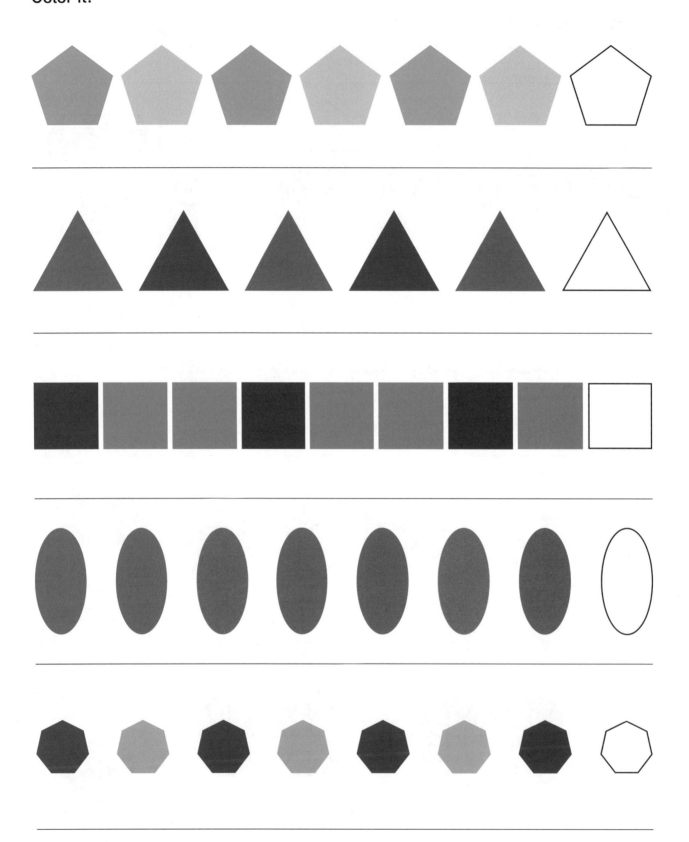

Using this page: Have students look at each pattern and color to continue the pattern correctly.

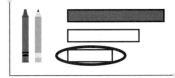

Color the longest and circle the shortest.

Using this page: Have students compare the objects, then color the longest and circle the shortest.

Review 9 Length

89

Circle the ruler that is longer than the blue one.

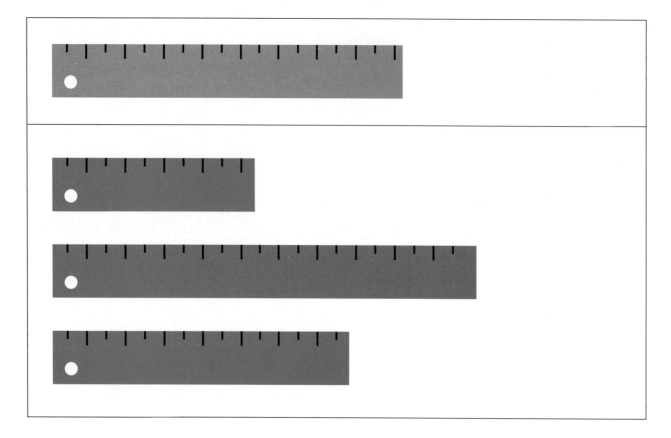

Circle the pencil that is shorter than the purple one.

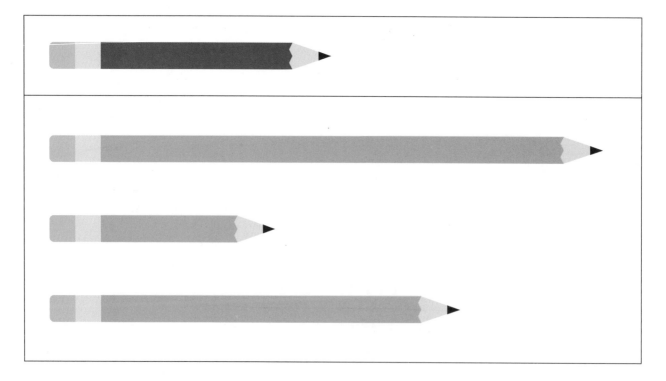

Using this page: Have students compare the objects to the one at the top and circle as specified.

90   Review 9 Length

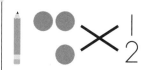

Exercise 10

Match.

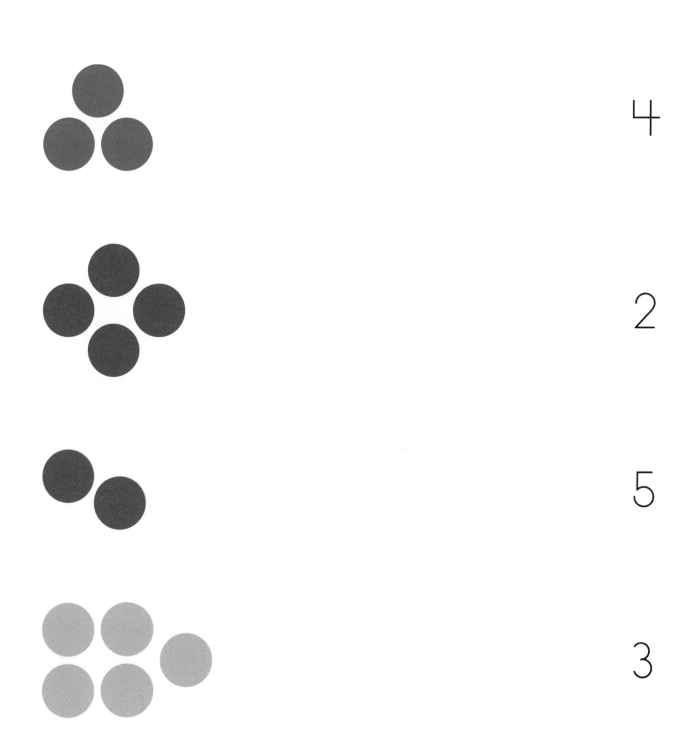

4

2

5

3

Using this page: Have students identify the number of dots quickly and match to the correct numeral.

Review 10 How Many?

91

Circle the correct number.

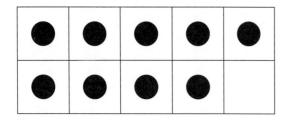

6   9   10

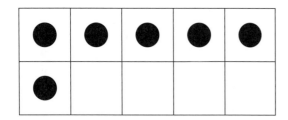

7   8   9

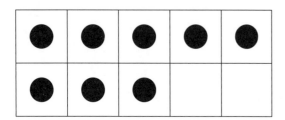

5   7   6

4   6   8

5   10   8

Using this page: Have students identify the number of dots and circle the correct numeral.

The shape with a face is first.

Color the third circle.

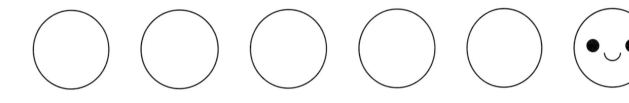

Color the second square.

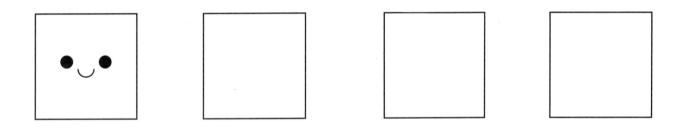

Color the fifth triangle.

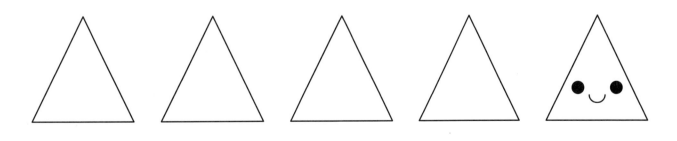

Using this page: Have students identify the shape with a face as first, then order the positions of the rest to find and color the specified shape.

The shape nearest to the flag is the first one.
Color the correct shape in each row.

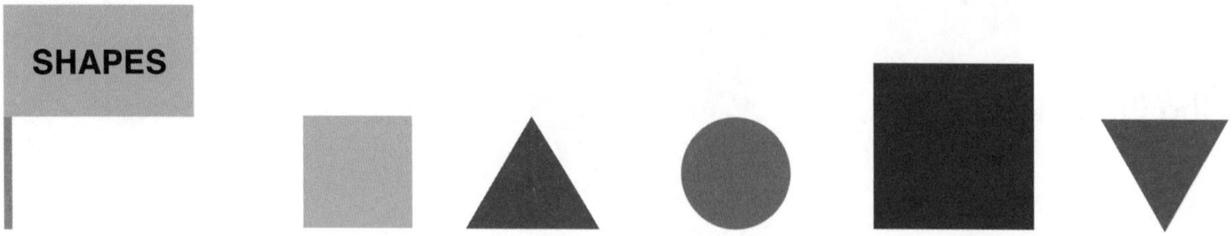

SHAPES

Fourth	□	△	○	□	▽
Second	□	△	○	□	▽
Third	□	△	○	□	▽
First	□	△	○	□	▽
Fifth	□	△	○	□	▽

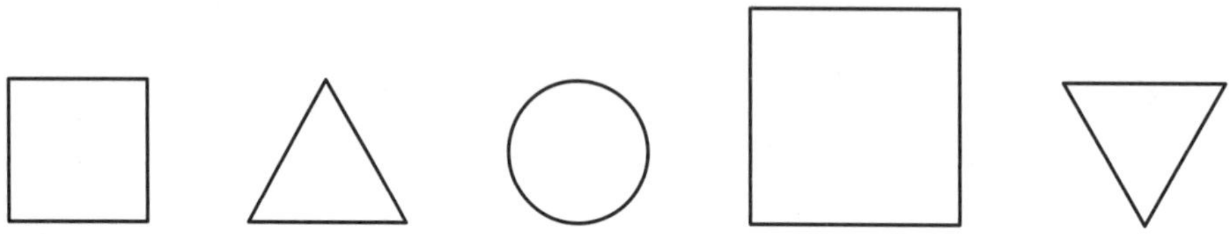

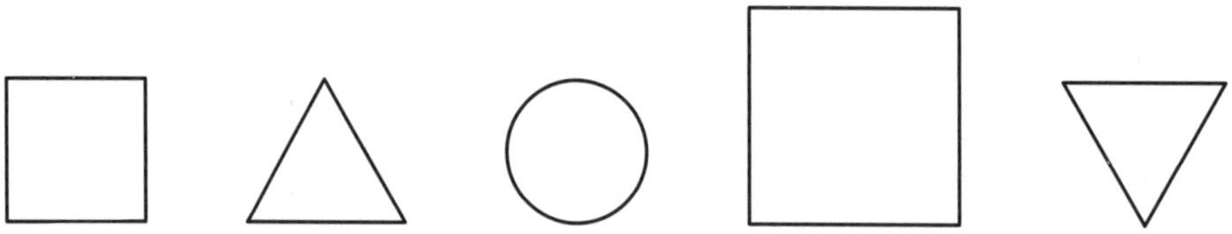

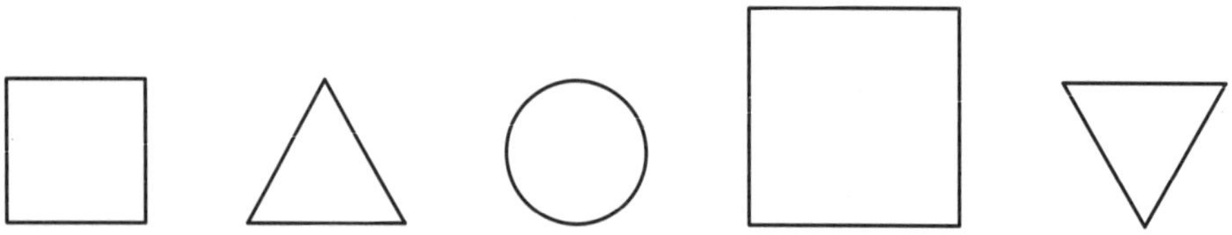

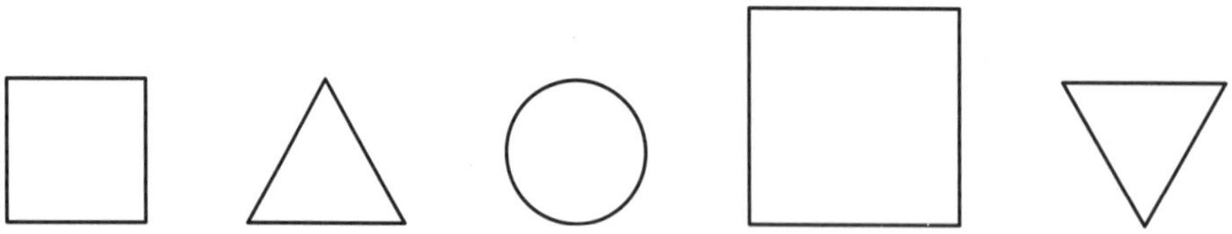

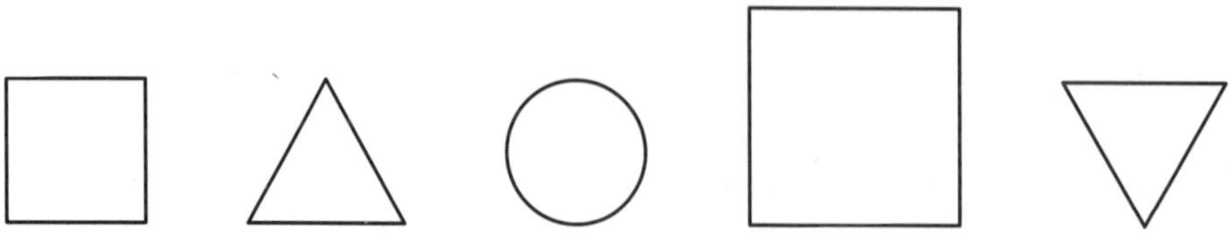

Using this page: Have students identify the flag as the front of the line of shapes. As you read each ordinal number, have students identify the shape in that position and color it the same color as shown above.

Exercise 12

Color according to the Color Key.

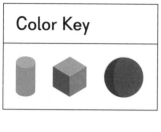

Color Key

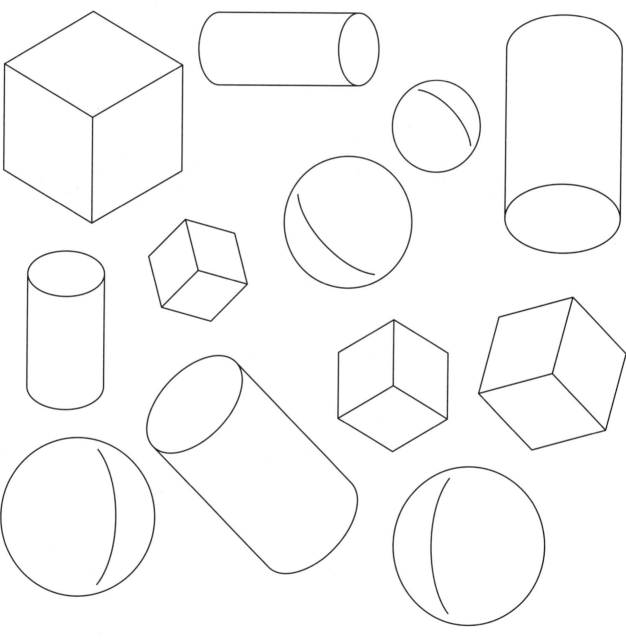

Using this page: Have students identify the solids and color them according to the color key.

What comes next?
Circle it.

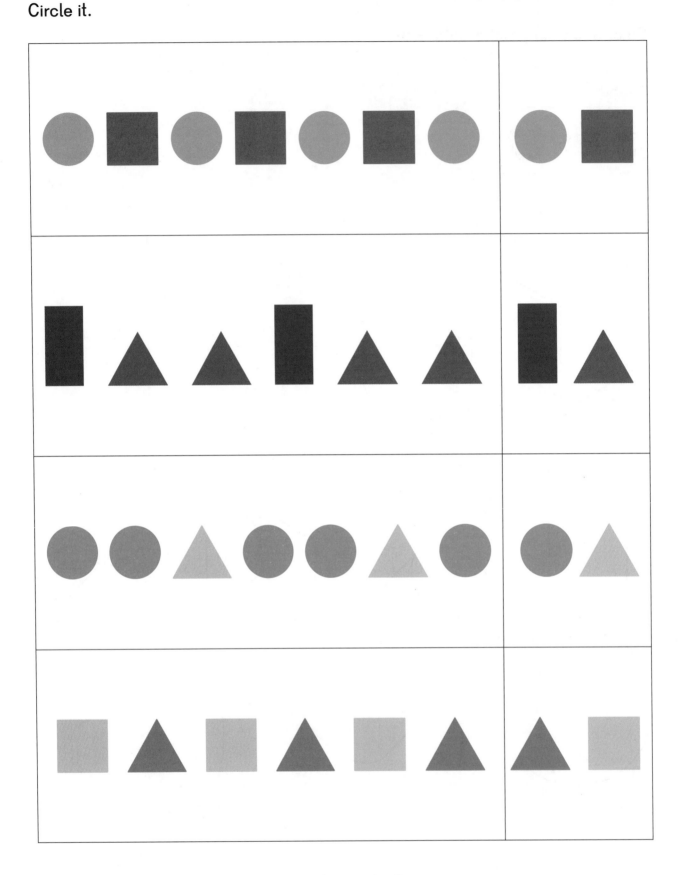

Using this page: Have students circle the shape that continues each pattern.

Review 12 Solids and Shapes

Which group has more?
Circle it.

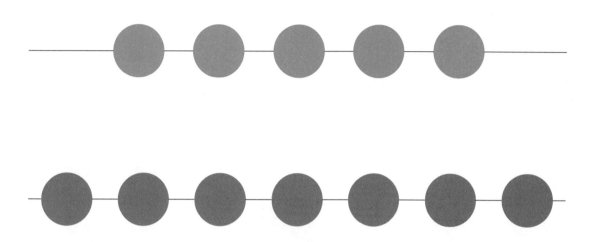

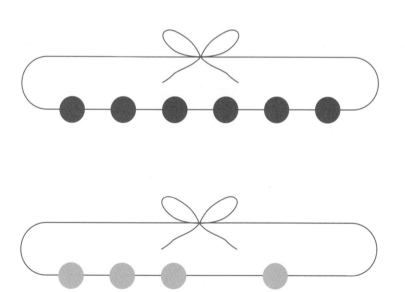

Using this page: Have students compare the two sets and circle the one that has more.

Match each  to a 🪏.
Are there more 🪏 or more 🪣?
Circle the answer in the box.

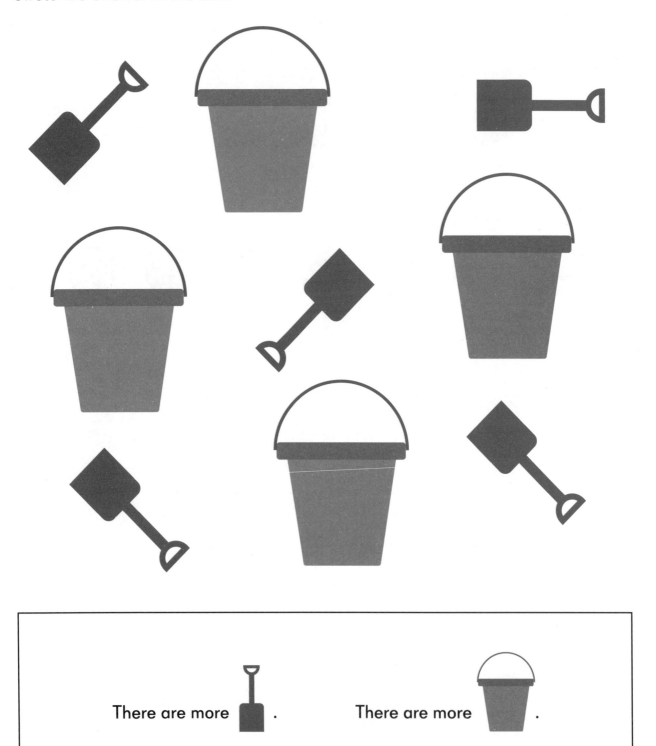

There are more 🪏.   There are more 🪣.

Using this page: Have students match the pails and shovels to find out which set has more, then circle the answer in the box.

Exercise 14

Which group has fewer?
Color that group.

Using this page: Have students compare the number of caps and hats, then color the set that has fewer.

Review 14 Which Set Has Fewer?   99

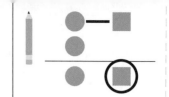

Match.

Are there fewer or ?

Circle the answer in the box.

There are fewer .   There are fewer .

Using this page: Have students match the ducklings and ducks to find out which set has fewer, then circle the answer in the box.

100   Review 14 Which Set Has Fewer?

Exercise 15

Circle to show how many altogether.

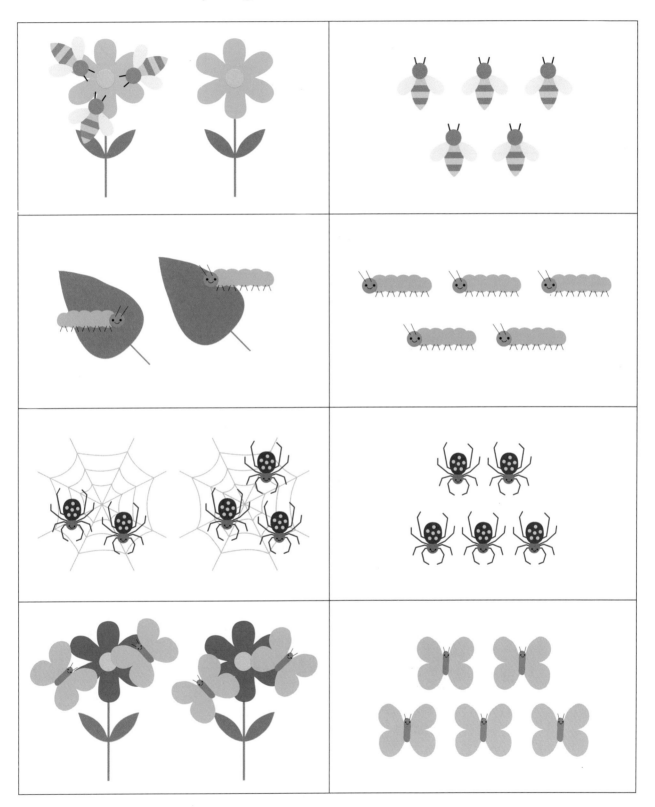

Using this page: Have students add the number of objects in the parts and circle that number of objects to show the whole.

Circle the number of ants in all.

 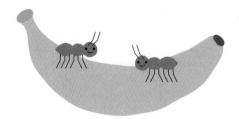   1   2   3

    3   5   2

 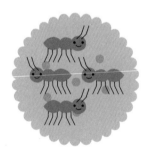   1   4   5

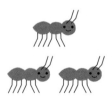    4   3   1

Using this page: Have students add the number of ants in the parts and circle the numeral to show the whole.

Exercise 16

Cross out the number of balloons that burst.
Circle the number left.

There were 5 balloons.
2 balloons burst.
How many balloons were left?

2   3   5

There were 3 balloons.
1 balloon burst.
How many balloons were left?

2   1   3

Using this page: As you read the number stories, have students cross out the specified number of balloons that burst to find the number of balloons left, then circle that numeral.

Circle the number left.

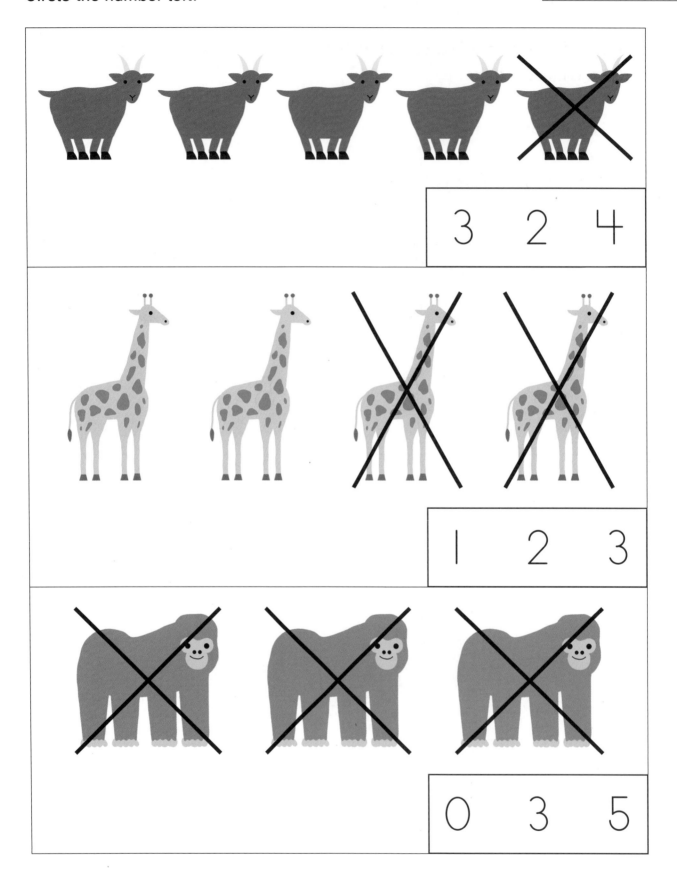

Using this page: Have students count the number of animals left, then circle the correct numeral.

Cut out these pictures for use on page 39 and 40.

Cut out these beads for use on page 41 and 42.

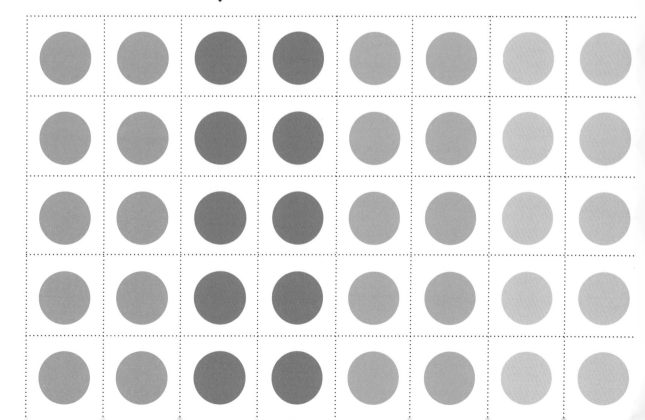

Blank